J. B. Madsen
G. E. Cold

The Effects of Anaesthetics upon Cerebral Circulation and Metabolism

Experimental and Clinical Studies

Springer-Verlag Wien GmbH

Jörn Bo Madsen, M.D.
Department of Anaesthesiology, Copenhagen County Hospital
at Hvidovre, Hvidovre, Denmark

Georg Emil Cold, M.D.
Department of Neuroanaesthesiology, Århus Kommunehospital,
Århus, Denmark

With 14 Figures

© 1990 by Springer-Verlag Wien
Originally published by Springer-Verlag Wien— New York in 1990
Softcover reprint of the hardcover 1st edition 1990

Typesetting: Macmillan India Ltd., Bangalore

Library of Congress Cataloging-in-Publication Data. Madsen, J. B. (Jörn Bo), 1951- . The effects of anaesthetics upon cerebral circulation and metabolism/J.B. Madsen, G.E. Cold.   p.   cm. Includes bibliographical references. ISBN 978-3-7091-3682-9   ISBN 978-3-7091-3680-5 (eBook) DOI 10.1007/978-3-7091-3680-5

1. Anesthetics—Physiological effect. 2. Cerebral circulation.
3. Brain—Metabolism. I. Cold, G.E. (Georg Emil), 1938-   II. Title. [DNLM: 1. Anesthetics—pharmacology, 2. Blood Flow Velocity—drug effects. 3. Brain—blood supply. 4. Brain—metabolism. 5. Cerebrovascular Circulation—drug effects. WL 302 M183e] RD85.5.M34 1990. 615'. 781—dc20. 90-10097

ISBN 978-3-7091-3682-9

Preface

The science of neuroanesthesia is fascinating and the amounts of experimental and clinical studies are overwhelming. Very early in our carreer as neuroanesthesiologists our memory concerning this field was challenged. Consequently, we initiated this review which also summarizes our clinical studies of the effects of anesthetics upon cerebral blood flow and metabolism.

The clinical studies were carried out at the Department of Anesthesiology, Hvidovre Hospital, Copenhagen, Denmark and the Department of Neuroanesthesia, Århus University Hospital, Denmark. Accordingly, we want to thank the chiefs of these departments, Professor Henning Ruben and Erland Hansen, M.D., Hvidovre Hospital and Paul Agerskov, M.D., Århus University Hospital. However, without the support by the chiefs of the respective neurosurgical departments it was impossible to accomplish the present clinical studies. We are grateful to Torkild Nørholm, M.D., Hvidovre Hospital and Professor Peter Rasmussen and Jens Buhl, M.D., Århus University Hospital.

We are deeply indebted to our collaborators in these clinical studies, both neurosurgeons and anesthesiologists, who have used their spare time and with enthusiasm have participated in the clinical studies. A special thank to the anesthesiologists Marianne Engberg, M.D., and Jan Asmussen, M.D., and the neurosurgeons Jarl Rosenørn, M.D., and Christian Kruse-Larsen, M.D. Furthermore, we want to thank Professor Jens Astrup, Department of Neurosurgery, Århus, who participated in several studies and advised us concerning the present publication and Professor Niels Lassen, Department of Neurophysiology, Bispebjerg Hospital, Copenhagen, who advised us concerning the clinical studies of cerebral blood flow by the Kety and Schmidt method.

This review is addressed to anesthesiologists, neuroanesthesiologists and neurosurgeons. It is the aim that the text might offer some help in the understanding of the effect of anesthesia upon brain blood flow and metabolism and inspire even deeper exploration into this interesting field.

Jörn Bo Madsen
Georg Emil Cold

Hvidovre and Århus, August 1990

Abbreviations

A	Adenosine
$AVDO_2$	Arterio-venous difference of oxygen
BBB	Blood–brain barrier
CA	Cerebral autoregulation
CBF	Cerebral blood flow
CBV	Cerebral blood volume
$CMRO_2$	Cerebral metabolic rate of oxygen
CPP	Cerebral perfusion pressure
CVP	Central venous pressure
CVR	Cerebral vascular resistance
CSF	Cerebrospinal fluid
HI	Head injury
ICP	Intracranial pressure
MABP	Mean arterial blood pressure
MCAO	Middle cerebral artery occlusion
MR	Magnetic resonance
NG	Nitroglycerin
SNP	Sodium nitroprusside
TMP	Trimethaphan

Contents

Introduction

During the last decade, the effects of anaesthetics on cerebral blood flow (CBF), cerebral metabolic rate of oxygen ($CMRO_2$) and intracranial pressure (ICP) have been studied experimentally and clinically, and several reviews have been published (Lassen and Christensen 1976, Frost 1984, Messick *et al.* 1985, Shapiro 1986). However, deductions concerning the effects of anaesthetics of cerebral circulation and metabolism, and the clinical applications of anaesthetics in neurosurgical practice are primarily based on experimental studies. One reason for this dilemma is, that clinical studies of CBF and metabolism during neuroanaesthesia is difficult because flow detectors have to be placed within the operative field.

In this review studies of CBF and $CMRO_2$ during craniotomy have been performed with the classical technique described by Kety and Schmidt (1945). Eight different anaesthetic techniques are presented including anaesthesia with the inhalation agents halothane, enflurane and isoflurane, anaesthesia with althesin, etomidate, neurolept anaesthesia and midazolam. In all studies the anaesthesia were supplemented with nitrous oxide 67%, fentanyl, and neuromuscular blockade was obtained with pancuronium.

This review consists of 9 chapters. In chapter 1 general considerations concerning the effects of anaesthetics on cerebral blood flow and metabolism are reviewed. In chapters 2 and 3, the effects of inhalation agents and hypnotics on flow and metabolism are considered. Chapters 4 and 5 cover the effects of central analgetics, and neuromuscular blocking agents. In chapter 6, the effects of other drugs in common use in neuroanaesthetic practice are summarized. Chapter 7 considers the effects of drugs used for controlled hypotension. In chapter 8, the application of Kety's method in studies of CBF and metabolism is reviewed, the studies of cerebral circulation and metabolism during nine different techniques of anaesthesia for craniotomy are presented, and other studies of cerebral circulation during neuroanaesthesia are reviewed. In chapter 9, considerations concerning central and cerebral hemodynamics during anaesthesia in the sitting position are considered.

This review, is primarily addressed to anaesthesiologists in the neuroanaesthetic clinic. Nevertheless, it is hoped, that this topic will also be of interest to those working within neurosurgery, neuroradiology and clinical neurophysiology.

1. General Considerations Concerning the Effects of Anaesthetics on Cerebral Blood Flow and Metabolism

With the exception of ketamine, and in experimental studies some of the synthetic analgetics (fentanyl, alfentanil and sufentanyl), all anaesthetics will induce a decrease in cerebral oxygen uptake. Concerning the hypnotic agents, including barbiturate, althesin, etomidate, benzodiazepines, neuroleptic agents and derivates of the fentiazines an associated decrease in CBF and $CMRO_2$ will occur. In contrast, inhalation agents like N_2O, halothane, enflurane and isoflurane generally give rise to a dissociation between CBF and oxygen uptake in the brain. These agents induce a fall in CVR, an increase in CBF and CBV, and a contemporary decrease in $CMRO_2$. The changes in ICP follow the changes in CBF and CBV. Thus inhalation agents will provoke an increase in CBF and ICP, and intravenous anaesthetics, with the exception of ketamine will result in a decrease in CBF and ICP.

During induction of anaesthesia with inhalation anaesthetics, the increase in ICP might be accompanied by a decrease in CPP, defined as the difference between MABP and ICP. This change is caused by the combined effect of myocardial depression and peripheral dilatation, which result in a decrease in MABP. In contrast intravenous anaesthetics generally keep the CPP unchanged, because the decrease in MABP is counterbalanced by a decrease in ICP.

Simultaneously with the decrease in $CMRO_2$, functional changes occur. Depending on the drug used, the dose of drug, the way of administration and the age of the patient, the subject will go through excitation, superficial unconsciousness until deep coma. These changes are accompanied by electro-encephalographic (EEG) changes, typically resulting in changes in frequencies from 14–20 Hz with low amplitudes to frequencies of 5–10 Hz with higher amplitudes. By increasing the dose of anaesthetic an enhancement of suppression with silent periods, disrupted by bursts of activity occur, until the EEG is totally isoelectrical.

In Table 1 the changes in CVR, CBF, $CMRO_2$, EEG, ICP, MABP and CPP are indicated after induction of anaesthesia with inhalation agents or bolus induction with intravenous hypnotics, analgetics and muscle relaxants. (Table 1 see p. 75.)

The maximal metabolic depression during normothermia after a bolus injection of barbiturate, althesin or etomidate is about 50% of the values obtained in the awake state. Thus, barbiturates given in a dose of

5–6 mg/kg will decrease $CMRO_2$ from the normal value of 3.5 ml O_2/100 g/min to about 1.7 ml O_2/100 g/min and decrease CBF from 50 ml/100 g/min to about 20–25 ml/100 g/min. A depression of $CMRO_2$ to these low values is accompanied by isoelectric EEG.

2. Inhalation Anaesthetics

Halothane (Experimental Studies)

In animal experiments halothane induces a decrease in CVR (Harp *et al.* 1976, Todd and Drummond 1984b), an increase in CBF (McDowall *et al.* 1963, Theye and Michenfelder 1968a, Smith and Wollman 1972, Smith 1973, Harp *et al.* 1976, Stullken *et al.* 1977, Todd and Drummond 1984b, Scheller *et al.* 1984), and a decrease in $CMRO_2$ (McDowall *et al.* 1963, Theye and Michenfelder 1968a, Smith 1973, Harp *et al.* 1976, Stullken *et al.* 1977, Todd and Drummond 1984b). The changes in CVR and CBF give rise to an increase in CBV and ICP (Fitch and McDowall 1971, DiGiovanni *et al.* 1974, Drummond *et al.* 1983a, Todd and Drummond 1984b). The increase in CBF is dependent on the age. Thus, the increase in CBF is more pronounced in young animals (Hoffman *et al.* 1982a). The increase in CBF occurs few minutes after induction, and is recorded before any change in $CMRO_2$ (Albrecht *et al.* 1977). During halothane administration over several hours a normalization of CBF occurs, and at the same time a moderate increase in CVR and $CMRO_2$ are observed (Albrecht *et al.* 1983). According to Warner *et al.* (1985), this cerebrovascular adaptation to prolonged halothane administration is not related to changes in CSF-pH, which is unchanged. Simultaneous administration of nitrous oxide give rise to an increase in CBF, and a decrease in CVR.

Halothane induces regional changes in CBF. Related to cardiac output a relative increase in rCBF has been observed in hypothalamus and in the brain-stem (Chen *et al.* 1982b), and it has been suggested that these regional differences in perfusion might play an important role in the regulation of the general circulation during halothane anaesthesia. Recent studies using an autoradiographic technique have unveiled regional differences in cerebral glucose utilization as well. Thus, CMR-glucose decreases relatively more in the occipital lobe, the brain-stem, cerebellar cortex and the anterior commissure (Shapiro *et al.* 1978).

The dissociation between CBF and $CMRO_2$ which occurs after halothane administration is dose dependent. Thus, an increase in the ratio $CBF/CMRO_2$, indicating a luxury perfusion has repeatedly been observed, and the same observation has been made during enflurane and isoflurane anaesthesia (Smith and Wollman 1972). However, the increase in this ratio during isoMAC administration of the three anaesthetic agents is most pronounced during halothane anaesthesia (Drummond and Todd 1985).

Halothane has a negative inotropic effect on the myocardium. This effect together with a decrease in peripheral systemic resistance, are responsible for the decrease in blood pressure. The net result is a decrease in CPP, partly caused by a decrease in MABP and partly owing to the increase in ICP (Jennett *et al.* 1967). The changes in CBF during halothane induction is dependent on CPP. Thus during gradual suppression of MABP, an associated decrease in CBF and $CMRO_2$ has been observed (McDowall *et al.* 1963), and during halothane induced hypotension to MABP 30 mm Hg a 50% decrease in CBF and $CMRO_2$ has been found. These changes are followed by a decrease in saturation of cerebral venous blood (Okuda *et al.* 1976, Miletich *et al.* 1976), but not by CSF-acidosis. Normalization of MABP after reduction of the inspiratory concentration of halothane induces an increase in CBF above control level. In this phase the cerebral autoregulation is impaired, and risk of cerebral edema is impending. Generally, the cerebral autoregulation is impaired or lost, dose dependently, during halothane anaesthesia (Okuda *et al.* 1976, Miletich *et al.* 1976, Morita *et al.* 1977, Albrecht *et al.* 1983), and this impairment is more pronounced during normo- and hypercapnia, but partly or completely normalized during hypocapnia (Miletich *et al.* 1976). Halothane induced impairment of the cerebral autoregulation is accompanied by a defect blood–brain barrier, and extravasation of protein and water in the extracellular fluid (Forster *et al.* 1978). In animal experiments of cryogenic lesion, halothane in comparison with barbiturates and neurolept anaesthesia induces an increase in the water content in cerebral tissue close to the lesion (Smith and Marque 1976). However, in recent studies with magnetic resonance scanning no difference in the water content has been disclosed (Fuller *et al.* 1980).

During halothane anaesthesia at normal CPP the CO_2 reactivity generally is preserved (Drummond and Todd 1985). However the CO_2 reactivity can be impaired during halothane induced hypotension (Okuda *et al.* 1976).

At high concentrations of halothane (5–10% inspiratory), a pronounced reduction of $CMRO_2$ has been found, and at these high concentrations an increase in lactate, the lactate/pyruvate ratio and a decrease in ATP and phosphocreatine in cerebral tissue occur (Michenfelder *et al.* 1970, Michenfelder and Theye 1975). In the brain cells, halothane inhibit GABA (gamma-aminobutyric acid), resulting in an increase in the concentration of GABA in the synapsis and inhibition of their function (Cheng *et al.* 1981). In connection with halothane anaesthesia, the concentration of GABA particularly increase in the pons (Fontenot *et al.* 1984). Halothane induces an increase in the electrical impedance of the brain, a result which is supposed to be caused by a reversible intracellular translocation of electrically, inactive ions from the extracellular space, possibly caused by an increased binding of cations to proteins (Schettini and Moreshead 1978).

Halothane gives rise to an increase in ICP (Fitch and McDowall 1971, DiGiovanni *et al*. 1974, Todd and Drummond 1984b). The increase in ICP is generally accompanied by a fall in CPP, and together these changes impede cerebral circulation, and augment the risks of cerebral ischaemia (Fitch and McDowall 1971). Experimentally an increase in intracranial pressure gradients between supra- and infratentorial compartments have been observed. These pressure gradients can increase considerably upto 50–60 mm Hg, and are followed by increased impaction at the tentorium and pupillary dilatation (Fitch and McDowall 1971).

Animal experiments and clinical research concerning the effect of halothane on cerebral circulation and ICP (vide infra), have to some extent resulted in avoidance of halothane in neuroanaesthesiological practice, especially in patients with space-occupying cerebral diseases (Jennett *et al*. 1969, Fitch and McDowall 1971, Gordon 1970). To some extent animal experiments support this view, especially because comparative studies of halothane, enflurane and isoflurane have shown that during IsoMAC concentrations of these anaesthetics, the increase in ICP is dose-dependent and most pronounced during halothane (Todd and Drummond 1984b). The same studies indicate that the increase in ICP is related to the increase in CBF. Furthermore, experimental studies in cats have shown that the tendency to brain prolapse during craniotomy is considerably higher during IsoMAC halothane anaesthesia (Drummond *et al*. 1983a).

Considering the increase in ICP effected by halothane, studies by Artru (1983) have shown that resistance to CSF resorption is increased, but CSF production is normal or decreased (Artru 1983d). The increased resistance to outflow is supposed to be a factor mediating sustained ICP increase after halothane anaesthesia (Artru 1983).

Human Investigations

Considering the experimental data concerning the effects of halothane on CBF and metabolism, studies in human support the findings in animal experiments. Thus, a decrease in CVR, a dose-related increase in CBF, and a decrease in $CMRO_2$ have repeatedly been observed (Alexander *et al*. 1964, Wollman *et al*. 1964, McHenry *et al*. 1964, Christensen *et al*. 1967). In the study by Christensen *et al*. (1967), 1% halothane in oxygen during normocapnia and maintained MABP increased CBF by 27% and decreased $CMRO_2$ by 26%. However, during normocapnic hypotension a 18% decrease in CBF was observed. These changes are supposed to be caused by impaired cerebral autoregulation as MABP support induced a considerable increase in CBF. In comparison IsoMAC studies of halothane, enflurane and isoflurane during normocapnia and MABP support have shown a dose related increase in CBF with all three anaesthetics. However, the increase in CBF during halothane already occurs during 0.6 MAC halothane, while the increase in CBF by enflurane and isoflurane are only registered with 1.6 MAC enflurane or isoflurane (Murphy *et al.* 1974). Studies during craniotomy using topical application of 133-Xe on the cerebral cortex close to cerebral tumors have supported these findings (Eintrei *et al.* 1985). The increase in CBF can induce critical intracranial hypertension, and decrease in CPP (Søndergard 1961, Marx *et al.* 1962, Jennett *et al.* 1967, Jennett *et al.* 1969, Shapiro *et al.* 1972a, Adams *et al.* 1972). These dangerous changes in ICP and CPP are frequently observed during induction of anaesthesia with halothane, especially during normo- and hypercapnia. In contrast, the increase in ICP to some extent can be prevented if hypocapnia is applied before administration of halothane (Adams *et al.* 1972). The attenuating effect of hypocapnia on ICP hyper- tension is most pronounced if hypocapnia is applied about 5 min before halothane administration (Misfeldt *et al.* 1974).

The effect of halothane on cerebral circulation has only been super- ficially investigated in infants. In preterm neonates without neurological diseases and subjected to halothane, isoflurane, fentanyl or ketamin anaesthesia, indirectly measured ICP decreases slightly (Friesen *et al.* 1987). The authors suggest that the difference between these results and those concerning studies of ICP in adults is due to compliance of the neonate's open-sutured cranium.

The CO_2 reactivity is preserved during halothane anaesthesia (Alexander *et al.* 1964), also when space-occupying lesions are present (Eintrei *et al.* 1985, Madsen *et al.* 1987a). The cerebral autoregulation is generally impaired in patients with space-occupying lesion (brain tumors, severe head injury etc). During craniotomy the autoregulation is especially challenged during incision and after extubation, where blood pressure increases can hardly be prevented. In studies of the arterio-venous oxygen content difference during anaesthesia with neurolept anaesthesia or an anaesthetic procedure with halothane 0.45–0.9% supplemented with ni- trous oxide as described by Madsen and Cold *et al.* (1987a) it has been shown that these two procedures generally provoke a decrease in $AVDO_2$ (Engberg *et al.* 1989), and clinically controlled studies with subcutaneous application of plain bupivacain in the scalp indicate, that the increase in MABP, and the decrease in $AVDO_2$ can be prevented by regional analgesia of the scalp before incision (Engberg *et al.* 1990).

During craniotomy for cerebral tumors in pentobarbitone induction, and maintenance of anaesthesia with fentanyl, nitrous oxide 67%, halo- thane 0.5% and moderate hypocapnia, an associated decrease in CBF and

$CMRO_2$, very like the decrease observed after intravenous bolus injection of barbiturates have been observed (Astrup et al. 1984). The same associated decrease has been observed after thiopentone induction, supplemented with nitrous oxide 67% and halothane 0.45% when hypocapnia has been applied. During these conditions only minimal increase in CBF has been found when the halothane concentration was increased from 0.45% to 0.9% (Madsen et al. 1987a). Considering the values of CBF and $CMRO_2$ obtained during the two investigations, they are identical to those obtained during neurolept anaesthesia (Cold et al. 1988). However, comparative studies of the arterio-venous oxygen content difference ($AVDO_2$) during halothane and neurolept anaesthesia with the same degree of hypocapnia, have shown that the $AVDO_2$ is maintained at lower levels during halothane anaesthesia. As $AVDO_2$ indicate the ratio $CBF/CMRO_2$, these dynamic studies are indirect evidence of the presence of a relative luxury perfusion elicited during halothane anaesthesia (Engberg et al. 1989). On Fig. 1 the dynamic changes in $AVDO_2$ during the two anaesthetic procedures are indicated.

As indicated before, animal experiments supported by human investigations have attenuated the disrepute of halothane in neurosurgical practice, especially in patients with space-occupying lesions (Jennett et al. 1969, Gordon 1970, Fitch and McDowall 1971). However, clinical studies of MABP, CPP, ICP, CBF and $CMRO_2$ during halothane anaesthesia show that halothane, like isoflurane, can under controlled circumstances and during the application of hypocapnia be used during neurosurgical intervention for small space-occupying processes (Astrup et al. 1984, Madsen et al. 1987a). As the CO_2 reactivity and the metabolic effect of thiopental in doses of 4–5 mg/kg under these circumstances are preserved, preoperative complications during this anaesthetic procedure such as brain prolapse or oedema can to some extent be controlled by intensifying the hypocapnia, or by intravenous injection of small doses of barbiturates (Astrup et al. 1984, Madsen et al. 1987a). Furthermore, clinical practice shows that mannitol in doses of 0.5–1 gr/kg or furosemide will produce shrinkage of the brain before or after opening of the dura.

Hyperventilation is generally advocated to patients subjected to neurosurgical procedures for space-occupying lesions. In clinical practice the lower limit of hypocapnia is supposed to be 2.7–3.3 kPa (20–25 mm Hg) (Gordon 1974), as animal experiments have shown that $PaCO_2$ below this limit does not reduce ICP further. On the contrary, the CPP might be reduced, and ICP increase if these limits are exceeded owing to an increase in CVP and a decrease in venous blood flow to the heart (Kitahata et al. 1971). In clinical practice moderate hypocapnia to $PaCO_2$ ranging from 3.6–4.0 is advocated (Frost 1984).

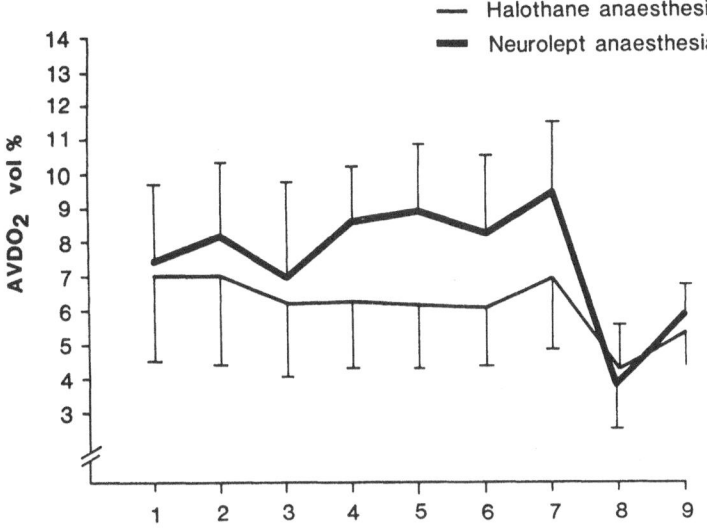

Fig. 1. Arterio-venous oxygen content difference ($AVDO_2$) measured repeatedly during two anaesthetic procedures in patients subjected to craniotomy for supratentorial cerebral tumors. Halothane anaesthesia included thiopentone induction, and maintenance of anaesthesia with halothane 0.5% in 67% nitrous oxide, and supplements with fentanyl. Neurolept anaesthesia included thiopentone induction, droperidol 0.2 mg/kg, fentanyl 4 µg/kg/h, nitrous oxide 67%. Both groups were hyperventilated to $PaCO_2$ levels about 4 kPa. *1*: Indicate $AVDO_2$ after placement of the jugular catheter, about 15 min after induction. *2*: before incision, *3*: 5 min after incision, *4*: after opening of the dura, *5*: after closure of the dura, *6*: before hyperventilation test, *7*: 5 min after 50% increase in pulmonary ventilation, *8*: 5 min after extubation, *9*: one hour after extubation (Acta Anaesthesiol Scand 1989: 33: 642–646, with permission)

As previously indicated, injection of thiopentone as a bolus dosis during halothane anaesthesia, will induce a further decrease in $CMRO_2$. Thus, thiopentone in doses of 4–6 mg/kg will reduce $CMRO_2$ to values of 1.7 ml O_2/100 g/min, which is close to the maximal suppression of $CMRO_2$ obtained by barbiturates alone (Astrup *et al.* 1984). On Fig. 2 the mean values of CBF and $CMRO_2$ in the awake state, during the first 3 hours of anaesthesia with halothane 0.5%, supplemented with nitrous oxide, fentanyl, and after thiopentone loads of 8 and 16 mg/kg are indicated. In comparison with the awake state, the anaesthesia is associated with an associated decrease of both CBF and $CMRO_2$. Thiopentone loads decrease flow and metabolism even further. As experimental studies have shown that barbiturates have a brain-protective effect, thiopentone supplemented general anaesthesia with halothane has been used in surgery for cerebral aneurysms (Bendtsen *et al.* 1984, Belapavlovic *et al.* 1985).

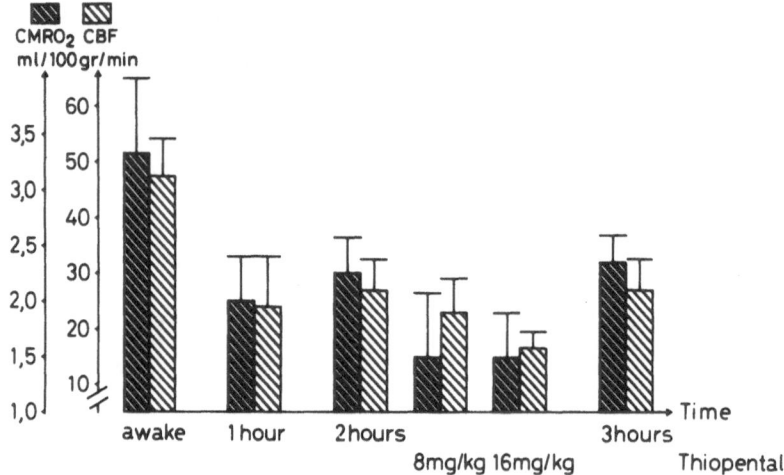

Fig. 2. CBF and $CMRO_2$ measured before and during anaesthesia with pentobarbitone induction, halothane 0.5% in nitrous oxide 67%. The awake state included 3 patients. The studies performed 1, 2 and 3 hours after induction included 6 patients, and the studies in which the patients were subjected to thiopentone 8 and 16 mg/kg included 5 patients in each group. All patients were studied during craniotomy for supratentorial tumors, and subjected to a $PaCO_2$ level averaging 3.6 kPa (Acta Anaesthesiol Scand 1984: 28: 478–481, with permission)

Clinically controlled studies concerning the beneficial effects of barbiturates have not been performed. Furthermore, barbiturate often given in doses of 10–20 mg/kg during craniotomy for aneurysms results in delayed postoperative recovery and postoperative respiratory insufficiency, which make postoperative artificial ventilation necessary (Bendtsen *et al.* 1984, Belapavlovic *et al.* 1985).

Conclusion

Repeated experimental as well as human studies have shown that halothane elicits a dose-related increase in CBV, CBF and ICP, and a decrease in CVR, $CMRO_2$, CMR-glucose and CPP. These results, combined with studies which indicate an impairment of cerebral autoregulation, and the brain–blood barrier, with risks of a development of cerebral edema, have discredited the use of halothane in surgery for space-occupying intracranial diseases.

The changes in cerebral circulation, are to some extent modified by the application of hypocapnia, barbiturate induction and a sufficient dosis of analgetics. The risks of intracranial hypertension and impeding decrease in

CPP, are related to the concentration of halothane used, the age of the patient and the haemodynamic stability. However, clinical studies in selected patients subjected to craniotomy for small space-occupying lesions like tumors, have shown that halothane in the concentration of 0.5–0.9%, combined with nitrous oxide and fentanyl can safely be used, and preoperative complications such as brain prolapse or edema can be controlled by the use of intensified hyperventilation, intravenous bolus dosis of thiopentone, and infusion of mannitol and furosemide.

Enflurane

Experimental Studies

In animal experiments enflurane induces a dose-dependent increase in CBF (Michenfelder and Cucchiara 1974, Takasaki 1974). However, if arterial pressure is not maintained, a fall in CBF has been observed (Sakabe 1975). $CMRO_2$ is decreased dose-dependently by enflurane (Stullken et al. 1977). At concentrations of enflurane above 2.2% in dogs a 34% decrease in $CMRO_2$ occurred, and at higher concentrations of enflurane, no further decrease in $CMRO_2$ was observed (Michenfelder and Cucchiara 1974). In rats 1.2 MAC combinations of nitrous oxide and enflurane produced a biphasic metabolic response. Local cerebral glucose utilization was activated in elective sensory input structures and components of the limbic system when nitrous oxide was increased from 0 to 30%, with a relative depression in metabolism when nitrous oxide was further increased from 30 to 60%. At all spinal cord levels a homogeneous increase in metabolism was observed when enflurane was replaced by 0–30% nitrous oxide change, with a return to control when nitrous oxide was further increased from 30–60% (Cole and Shapiro 1989).

At normal ICP in dogs, enflurane only increases ICP slightly if at all (Boop and Knight 1978, Hans et al. 1980). However, a significant increase in ICP was observed in animals with intracranial hypertension (Boop and Knight 1978), and the increase in ICP was smaller compared with halothane (Hans et al. 1980). Moreover, in cats the degree of brain protrusion found during craniotomy was more pronounced during halothane, compared to enflurane and isoflurane anaesthesia (Drummond et al. 1983a). After discontinuing the administration of enflurane a significant increase in ICP without correlation to changes in CBV has been observed experimentally (Artru 1983). These changes are supposed to be caused by an increase in resistance to CSF reabsorption and in the production rate of CSF during enflurane anaesthesia (Mann et al. 1980, Artru 1984c).

During the first experimental studies with enflurane, twitching of the extremities and face were observed (Virtue *et al.* 1966, Botty *et al.* 1968). Later on, de Jong and Heavner (1971), found that the twitching was synchronous with spike discharge activity obtained by EEG, and Joas *et al.* (1971) studying dogs subjected to hypocapnic enflurane anaesthesia, observed seizure-like EEG tracings and focal twitching in response to hand clapping, and disappearance of spike activity when $PaCO_2$ was increased. In cats, there is a prolonged period of abnormal EEG activity of up to 16 days following enflurane anaesthesia (Julien and Kavan 1972). In the same animal, diazepam, thiopentone, methohexitone and ketamine all enhanced established EEG seizure activity (Darimont and Jenkins 1977).

In dogs the seizure activity associated with enflurane anaesthesia is accompanied by a 48% increase in $CMRO_2$ (Michenfelder and Cucchiara 1974). The hyperexcitability appears to originate in the limbic system, with subsequent spread to all other areas (Darimont and Jenkins 1977). Myers and Shapiro (1979) have suggested that the enflurane induced epileptogenic foci are located in the hippocampus and related structures, and from the same group a significant decrease in the ratio rCBF/CMR-glucose in the hippocampus was observed, suggesting that substrate demand could potentially outstrip supply within the hippocampus (Ray *et al.* 1979). However, other studies in rats have shown that the cerebral cortical energy state is not depleted during enflurane induced seizure activity (Seo *et al.* 1984), and neuropathological studies in rats, subjected to long-time enflurane EEG spiking activity have not shown any signs of irreversible neuronal damage (Lin *et al.* 1986).

Clinical Studies

In normal patients enflurane has no effect on global CBF (Wollman *et al.* 1969, van Aken *et al.* 1977), but a significant decrease in regional CBF in the frontal and occipital regions at 2% enflurane has been observed (Rolly and van Aken 1979). In patients with subacute or chronic brain lesions, enflurane causes a decrease in CBF (Reinhold *et al.* 1974, Reinhold and DeRood 1976), and in patients with head injury whose blood pressure was maintained with phenylephedrine, 1% enflurane had no effect on global CBF (De Rood *et al.* 1980). However, in many of these studies, the patients were premedicated, and anaesthesia was maintained with nitrous oxide. Consequently, Sakabe *et al.* (1983) compared pure enflurane anaesthesia at two concentrations with values obtained at awake state, and found that enflurane at 2 and 3.5% is a cerebral vasodilator causing a 15 and 32% increase in CBF respectively. In comparative clinical studies of enflurane, halothane and isoflurane isoMAC concentrations of isoflurane and enflurane give rise to a comparable increase in CBF, which is much smaller than the increase obtained with halothane (Murphy *et al.* 1974, Eintrei *et al.* 1985).

Enflurane gives rise to a decrease in $CMRO_2$ (Wollman *et al.* 1969, Sakabe *et al.* 1983). In patients with supratentorial cerebral tumors, subjected to craniotomy with thiopentone induction and maintenance of anaesthesia with nitrous oxide 67%, fentanyl and enflurane 1 or 2% during hypocapnia, an associated decrease in CBF and $CMRO_2$, and a dose related decrease in $CMRO_2$ have been observed (Madsen *et al.* 1986). Compared to IsoMac concentrations of isoflurane and halothane, the suppression of $CMRO_2$ is more pronounced, and the effect on global CBF identical (Madsen and Cold 1987).

Induction of anaesthesia with 3% enflurane in patients with intracranial lesions gives rise to a marked increase in ICP (Tambuniello *et al.* 1978). In contrast, enflurane 1% during induction in high-risk neurosurgical patients does not change ICP (McLeskey *et al.* 1974), and in patients with cerebral tumors, enflurane 2% has very little effect on ICP, but decreases blood pressure and CPP significantly (Moss *et al.* 1983). During induction, the increase in ICP with enflurane is similar to that obtained with halothane (Stullken and Sokoll 1975). In patients with intracranial lesions and high ICP level, enflurane 1–2% increases ICP by up to 9–15 mm Hg (Zattoni *et al.* 1974), and in patients with severe head injury, having an initial ICP above 20 mm Hg, enflurane 1.5% induces a considerable increase in ICP, similar to nitrous oxide and halothane (Schulte am Esch *et al.* 1979). However, in other studies the ICP increase elicited by enflurane was smaller than that obtained with halothane (Cunitz *et al.* 1976).

In many of these studies, enflurane decreases CPP significantly (Zattoni *et al.* 1974, Cunitz *et al.* 1976, Schulte am Esch *et al.* 1979, Moss *et al.* 1983) often to values less than 50 mm Hg, which is at the lower end of the range of CA. Thus, changes in CA capacity, and the fact that most studies in patients with presumably lost CA, might influence the changes in ICP, and it is supposed that the increase in ICP during enflurane anaesthesia might be even more pronounced if blood pressure is maintained.

Human studies have confirmed that epileptic activity might be elicited by enflurane. The incidence of muscle twitching in patients receiving enflurane has been reported to be 7% (Lebowitz *et al.* 1972), and in children enflurane consistently produced spike waves on EEG even at concentrations as low as 1% (Neundörfer and Klose 1975). Increasing depth of enflurane anaesthesia is characterized by the appearance of high-voltage spikes (Stockard and Bickford 1975), but with elevation of $PaCO_2$, indication of cerebral irritability appears to be reduced; in the same study, the addition of nitrous oxide did not alter the EEG patterns (Neigh *et al.* 1971). The EEG changes can persist for several days (Burchiel *et al.* 1977, Kruczek *et al.* 1980), and epileptic seizures have been observed days after enflurane administration (Ohm *et al.* 1975, Grant 1986). In a study of 14 patients subjected to craniotomy for small cerebral tumors, the combined use of hypocapnia and enflurane in concentrations of 1–2% did not elicit epileptic seizure activity (Madsen *et al.* 1986). For review, see Steen and Michenfelder (1979).

Enflurane increases the latency of brain-stem auditory evoked potential response in human (Dubois *et al.* 1982, Thornton *et al.* 1983). However, inter-peak latencies are also affected, and the abnormalities are suggestive of a maximal effect of enflurane on midbrain reticular formation. These observations seem to confirm the hypothesis of the rostral brain-stem playing a role in the mechanism of generalized seizure activity (Mirsky *et al.* 1979).

Conclusion

Experimental as well as human studies have shown that enflurane is a cerebral vasodilator which gives rise to an increase in CBF and ICP, and a decrease in CVR. Simultaneously enflurane induces a marked decrease in $CMRO_2$. Owing to the phenomenon of epileptic seizure activity observed during enflurane anaesthesia, this inhalation agent is generally avoided in the neurosurgical clinic. However, provided EEG monitoring can be practised pre- and postoperatively, clinical studies suggest that enflurane might be an alternative to halothane (Madsen *et al.* 1986).

Isoflurane

During the last decade, experimental and clinical studies concerning the effect of isoflurane have been innumerable and several reviews have been published (Eger II 1981, Eger II 1984, Frost 1984).

Experimental Studies

In animal experiments, isoflurane induces an increase in CBF and a decrease in CVR, simultaneously with a decrease in $CMRO_2$ (Cucchiara *et al.* 1974, Stullken *et al.* 1977, Newberg *et al.* 1983, Todd and Drummond 1984b, Gelman *et al.* 1984). The changes in CBF and $CMRO_2$ give rise to an increase in the ratio $CBF/CMRO_2$ (Smith and Wollman 1972). In dogs, isoflurane, even during hypocapnia may increase CBV (Archer et al. 1987). However in rats subjected to 1 MAC isoflurane or halothane there was no difference in CBV (Katz et al. 1988). The increase in CBF occurs fastest and more pronounced in the deep structures of colliculus superior and inferior, the limbic system and the cerebellum, as compared with the cerebral cortex (Maekawa et al. 1983, Manohar and Parks 1984, Mutch and Ringaert 1987). Comparative experimental studies in cats and dogs indicate that isoflurane possesses cerebrovascular properties that are different from halothane. Thus, the increase in CBF and decrease in CVR elicited by isoflurane is much smaller (Todd and Drummond 1984b), and rCBF to cortical and subcortical areas, including diencephalon and the brain stem, were all lower during isoflurane than halothane anaesthesia (Chen et al. 1984). Studies of CBF by hydrogen clearance in rabbits subjected to hypo-, normo- and hypercapnia support these findings. Thus, CBF during 1 MAC isoflurane decreased during hypocapnia, was unchanged during normocapnia and increased regionally during hypercapnia. In comparison, during 1 MAC halothane CBF increased during all three conditions (Scheller et al. 1986a). However, discrepancies exist as to the degree of vasodilatation effected by the two agents. Recently, Hansen et al. (1988a) studied CBF autoradiographically in rats, and found that cortical CBF was greater during halothane anaesthesia, while subcortical CBF was greater during isoflurane anaesthesia. The reason for these anatomically selective CBF effects by the two agents is unknown, but is supposed to be related to their different cortical electrophysiological effects.

Unlike halothane, where a decrease in CBF is observed during prolonged administration, CBF and $CMRO_2$ during isoflurane anaesthesia of 3–4 hours duration remain unchanged, and the EEG recordings show a continuous sleep pattern (Roald et al. 1989).

In studying the direct cerebral vasodilating potencies of halothane and isoflurane in rabbits Drummond et al. (1986) found that the relative effect of halothane and isoflurane on CBF is dependent on the $CMRO_2$ present prior to their administration. When the preexisting $CMRO_2$ is not markedly depressed, isoflurane decreases $CMRO_2$ and causes less cerebral vasodilatation than does halothane. However, if the initial $CMRO_2$ is depressed by a high dose of pentobarbitone, which caused a 43% decrease in $CMRO_2$, halothane and isoflurane have similar vasodilating potencies. However, other studies in the same animal indicate that pentobarbitone given in a bolus dosis during isoflurane anaesthesia until burst suppression on EEG, will increase CBF (Scheller et al. 1987a).

The effect of isoflurane anaesthesia on local cerebral glucose utilization has been studied in rats. A significant decrease was observed in all cortical

areas and in the primary sensory relay nuclei of the central visual and auditory pathways. Furthermore, the utilization of glucose was decreased in
the cerebellum, red nucleus, the ventral thalamus, in the CA_1 field and
dentate gyrus in the hippocampus. However, the utilization was increased
in the substantia nigra pars compacta and in the medial habenulo-interpen-
ducular system and the CA_3 field of the hippocampus (Ori *et al.* 1986,
Maekawa *et al.* 1986). During isoflurane anaesthesia in rats, hypocapnia to
$PaCO_2$ 20 mm Hg at normotension does not change glucose utilization,
while rCBF becomes homogeneously reduced by 40%. On the other hand,
hypocapnia combined with isoflurane-induced hypotension to MABP
50 mm Hg causes a heterogenous response with CMR-glucose decrease in
the frontal and the parietal cortex and the thalamus, unaltered utilization in
the mesencephalon and the medulla oblongata, and increased utilization
in the hippocampus and the cerebellum (Waaben *et al.* 1989). In comparative studies in rats during 1 MAC isoflurane or halothane a strong correlation between glucose utilization and CBF within individual anatomic
regions was found. Furthermore, at a given value for glucose utilization
CBF during isoflurane was higher than with halothane suggesting that
CBF and oxygen consumption remain coupled, and that at a given level of
glucose utilization, isoflurane possesses greater cerebral vasodilating cap-
abilities than halothane (Hansen *et al.* 1988b).

In dogs, the benzodiazepine antagonist, flumazenil, partially antagonizes the suppression of $CMRO_2$, and EEG is changed from sleep to awake
pattern (Roald *et al.* 1988).

Comparative studies of CA during isoflurane and halothane anaesthesia
indicate an impairment of CA during halothane, while CA is partly preserved during isoflurane anaesthesia (Todd and Drummond 1984b). However, during isoflurane-induced hypotension in baboons, the CA is totally
abolished, and after normalization of blood pressure, secondary to the
reduction of the isoflurane concentration, a short lasting increase in CBF
has been observed, indicating a hyperaemic phase (Van Aken *et al.* 1986).
Recent studies in dogs by McPherson and Traystman indicate, that during
decrease in MABP or increase in CSF pressure the CA is preserved at
1 MAC isoflurane, but absent at 2 MAC.

In animals without cerebral lesions and blood pressure within a normal
range of CA, the CO_2 reactivity is preserved (Cucchiara *et al.* 1974), and the
reactivity is higher with 1 MAC isoflurane compared with 1 MAC halothane (Drummond and Todd 1985). In baboons subjected to isoflurane-induced hypotension the CO_2 reactivity was found to be intact at
base-line blood pressure; however, at 20 and 50% decrease in blood
pressure the CO_2 reactivity was found to be impaired and absent respectively (Van Aken *et al.* 1986). Other studies in dogs subjected to 1.4%

isoflurane indicate preserved CO_2 reactivity to hypo- as well as hypercapnia; however with 2.8% isoflurane vasoconstriction to hypocapnia is retained, but vasodilation to hypercapnia is abolished (McPherson *et al.* 1989).

In studies in cats and dogs, isoflurane only increases ICP minimally if at all (Todd and Drummond 1984b). The same results were found in dogs with intracranial hypertension caused by inflation of an epidural balloon and subjected to 1 MAC isoflurane; however, if nitrous oxide was supplemented to isoflurane, an increase in ICP was observed (Albin *et al.* 1986). It is supposed that species differences might exist, because Scheller *et al.* (1987) in rabbits subjected to cryogenic brain lesion observed an increase in ICP independent of the level of $PaCO_2$. In cats subjected to craniotomy, brain surface protrusion during enflurane and isoflurane anaesthesia is much less than during halothane anaesthesia, and when anaesthetic-induced differences in blood pressure were eliminated by arterial pressure support, the protrusion caused by halothane as compared with that caused by enflurane or isoflurane was exaggerated (Drummond *et al.* 1983a).

During prolonged isoflurane anaesthesia in dogs, CBF declines over a period of hours despite constant CPP (Brian *et al.* 1988). In other studies the increase in ICP during isoflurane administration seems to follow changes in CBV, which is increased by 10%. However, the increase in ICP is only of 30 min duration as opposed to more than three hours duration with halothane and enflurane (Artru 1984a). In dogs, isoflurane in comparison with enflurane does not change the rate of CSF production (Artru 1984b).

Isoflurane causes progressive changes in the EEG. At concentration less than 0.5 MAC an increase in alpha frequency from 8–12 Hz to more than 15 Hz occurs; voltage increases concomitantly. When isoflurane concentration is increased above 0.4 MAC, the regions with high voltages shift from the posterior to the anterior part of the brain. At 1 MAC, frequency decreases and voltage increases further. At 1.5 MAC, burst suppression appears, and the voltage begins to diminish. At 2 MAC, the EEG becomes isoelectric (Stullken *et al.* 1977, Tinker *et al.* 1977, Newberg *et al.* 1983).

Studies in mice subjected to hypoxia have shown that isoflurane has a brain-protective effect (Newberg and Michenfelder 1983). In dogs isoflurane-induced hypotension to MABP 40 to 50 mm Hg produced a 40% decrease in $CMRO_2$, accompanied by a 60% decrease in CBF while the cerebral energy state indicated by the concentrations of ATP, ADP, AMP, phosphocreatine and lactate were unchanged (Newberg *et al.* 1984). During exposure to isoflurane in concentrations of 5–6%, an increase in the concentrations of lactate and lactate/pyruvate ratio, but an unchanged concentration of ATP and phosphocreatine in cerebral tissue has been found (Newberg *et al.* 1983). The brain protective effect of isoflurane has been evaluated in several studies. With the exception of one study where isoflurane 2% showed significantly better protection than the protection found during nitrous oxide anaesthesia, and that the outcomes of the animals subjected to either isoflurane or methohexital were comparable (Baughman *et al.* 1989b), other experimental results unfortunately have been negative. Thus, Warner *et al.* (1986) studied the effect of isoflurane on neuronal necrosis following near-complete forebrain ischaemia in the rat (bilateral carotid occlusion and hypotension to 50 mm Hg), and found morbidity and pathophysiological changes unaffected by isoflurane; and studies in rats subjected to MCAO during controlled hypotension with either sodium nitroprusside, isoflurane or controlled hypovolaemia did not disclose any difference in zero flow areas as evaluated by the 14C-antipyrine technique (Cole *et al.* 1987). Studies in monkeys subjected the MCAO during thiopentone or isoflurane induced EEG suppression until burst suppression level, and where MABP was maintained with pressor support indicate that isoflurane does not offer advantages compared with thiopentone as regards cerebral protection (Nehls *et al.* 1987, Milde *et al.* 1988). These results are supported by *in vitro* studies of evoked population spikes recorded from brain hippocampus slices, where isoflurane compared with thiopentone did not elicit any brain protective effect (Bendo *et al.* 1987). Thus, it is still unknown as to what extent isoflurane offers any brain protective effects in regional ischaemia, and if so, whether isoflurane and thiopentone during strictly comparative circumstances offer the same degree of protection (Michenfelder 1987). However, the discrepancies in results concerning the brain-protective effect of isoflurane in focal ischaemia might be influenced by the brain glucose concentration, as a high brain concentration of glucose during incomplete ischaemia might worsen the outcome. That this factor might play a role in studies of isoflurane-induced brain protection, has indirectly been shown by Kofke *et al.* (1986), who found that brain concentration of glucose during isoflurane anaesthesia was higher than during halothane, suggesting that cerebral lactic-acidosis might be increased if isoflurane is given. Theoretically, nitrous oxide (an agent with cerebral stimulatory effect), might attenuate the cerebral protective effect of isoflurane. However, mortality studies in rats suggest that isoflurane provides better cerebral protection alone, than nitrous oxide, but does not indicate that nitrous oxide combined with isoflurane worsens the outcome (Baughman *et al.* 1988). In a later study the same group found that during moderate ischaemia 0.5 and 1 MAC isoflurane produced better neurologic outcome and less histopathologic damage from ischaemia than did nitrous oxide 70% alone. Addition of nitrous oxide to 0.5 MAC isoflurane attenuated the improvement in neurologic outcome and produced more histopathologic damage compared with that produced during isoflurane alone. However, the addition of nitrous oxide to 1 MAC isoflurane did not significantly change outcome or histopathology score (Baughman *et al.* 1989c).

In comparative studies in dogs of controlled hypotension to MABP 40 mm Hg with trimethaphan, nitroprusside, halothane or isoflurane, the energy charge in cerebral tissue was found to be unchanged during isoflurane-induced hypotension. In the same study CBF and $CMRO_2$ were significantly decreased during isoflurane-induced hypotension (Newberg *et al.* 1984). Comparative studies of pO_2 in cerebral tissue during isoflurane and 2-chloroadenosine induced hypotension indicate higher pO_2 levels during isoflurane-induced hypotension (Seyde and Longnecker 1986). Furthermore, studies in dogs subjected to isoflurane-induced hypotension to 40 mm Hg of MABP and hypocapnia ($PaCO_2$ 20 mm Hg) suggest no adverse effects on cerebral metabolism or function (Artru 1986a). The responses of CBF, $CMRO_2$, and ICP during varying rates of isoflurane-induced hypotension have recently been studied in dogs. On the one hand, slow induction of hypotension to MABP 45 mm Hg within 6–10 min in comparison with fast induction within 1–5 min resulted in a transient increase in ICP, and a more prolonged ICP increase after reestablishment of normotension and the authors conclude that the increase in ICP could be deleterious when intracranial compliance is already jeopardized (Hickey *et al.* 1986b). On the other hand, rapid induction of hypotension in comparison with slow induction results in a transient reduction of CBF that is not accompanied by a similar reduction in $CMRO_2$, suggesting that isoflurane-induced hypotension should be initiated gradually to permit enactment of homeostatic mechanisms that result in the least alteration in CBF (Hickey *et al.* 1986a). The comparative studies combined with experimental evidence of a cerebral protective effect of isoflurane has focused attention on isoflurane as an agent used in clinical practice for controlled hypotension.

Studies concerning the elimination of isoflurane from the brain are conflicting. In studies by Wyrwicz *et al.* (1983), a 43% decrease of the 19-F NMR signal was still observed 5 hours after a 90 min isoflurane anaesthesia. However, this result, indicating a very slow elimination of isoflurane disagrees with invasive results with end-tidal gas chromatography (Cohen *et al.* 1972, Carpenter *et al.* 1986). Recently NMR studies in which signals from extracerebral compartments were eliminated have shown a half-life of isoflurane elimination of 36 min (Mills *et al.* 1987).

Human Investigations

Comparative clinical studies in patients without cerebral lesions (Murphy et al. 1974, Algotsson et al. 1988), and in patients with mass-expanding cerebral lesions (Eintrei et al. 1985, Madsen and Cold 1987) have shown that the increase in CBF, if present at all, is less with iso-MAC isoflurane and enflurane than with halothane and that the increase in CBF is dose-dependent. Simultaneously, isoflurane induces a decrease in $CMRO_2$ (Murkin et al. 1986, Madsen et al. 1987b, Algotsson et al. 1988). In preoperative studies of patients subjected to craniotomy for small supratentorial tumors with thiopentone induction, and maintenance of anaesthesia with isoflurane 0.75%, nitrous oxide 67%, fentanyl with moderate hypocapnia, CBF is decreased by 27%, and $CMRO_2$ by 29% in comparison with the values obtained in the awake state. Under these circumstances, an increase in isoflurane concentration to 1.5% does not change CBF, while CVR and $CMRO_2$ show a further decrease. In the same study the relative CO_2 reactivity was found to be intact (Madsen et al. 1987b). Other studies indicate that the CO_2 reactivity in edematous brain tissue might be absent during isoflurane anaesthesia, but preserved during fentanyl anaesthesia (Shah et al. 1988).

In patients without intracranial diseases undergoing a coronary by-pass operation, 1 MAC isoflurane or enflurane at normothermia produced a 35 and 39% fall in CBF respectively, and $CMRO_2$ was decreased by 50% with either agents. The CA was impaired with both agents, while the CO_2 reactivity was maintained at 1 MAC (Larsen et al. 1988). During cardiopulmonary by-pass for closed heart surgery, the combination of hypothermia to about 26 C, and isoflurane until burst suppression level of EEG reduced CBF to levels of 10 ml/100 g/min, and $CMRO_2$ to about 0.3 ml O_2/100 g/min. Thiopentone infusion used to obtain the same level of EEG suppression will under the same circumstances reduce $CMRO_2$ to the same level, but reduces CBF even further until the level of 8 ml/100 g/min (Woodcock et al. 1987).

In preterm neonates, anterior fontanel pressure recorded with the Ladd AFP monitor, decreases by 11% during administration of 0.75% isoflurane (Friesen et al. 1987).

Generally isoflurane elicits a decrease in MABP. Studies of the central hemodynamics have shown that the decrease in MABP is primarily caused by peripheral vasodilatation and not a fall in cardiac output (Stevens et al. 1971, Shimosato et al. 1982).

In clinical studies isoflurane increases ICP and decreases MABP and CPP during normocapnia. However, the increase in ICP and the decrease in CPP can be diminished if hyperventilation or barbiturate injections are applied prior to or simultaneously with the administration of isoflurane (Adams *et al.* 1981, Campkin 1984, Saintz *et al.* 1988, Campkin and Flinn 1989). During neuroleptic anaesthesia for tumor and aneurysm surgery, isoflurane 1% increases ICP and reduces MABP and CPP, but controlled hyperventilation partially restores the ICP and CPP values (Gordon *et al.* 1988). In other studies where the patients were premedicated with droperidol and fentanyl, isoflurane 1% during normocapnia did not increase ICP, while 1% isoflurane during hypocapnia was associated with a decrease in ICP (Mazzarella *et al.* 1986). It is supposed that the discrepancy in results between this study and the study by Adams and coworkers, where ICP increased during normocapnic isoflurane administration may be due to differences in the regime of premedication. In other studies of patients with space-occupying cerebral lesions, an increase in ICP has also been observed (Grosslight *et al.* 1985, Belopavlovic and Buchthal 1986, Pfeifer *et al.* 1987), and caution in using isoflurane in patients with limited intracranial compliance has been advanced.

In clinical studies, isoflurane administration is followed by suppression of the EEG activity just like the suppression obtained in experimental studies (Eger *et al.* 1971, Homi *et al.* 1972, Clark *et al.* 1973), and at concentrations of 2–3% isoflurane, the EEG is isoelectric. These changes are not observed during halothane anaesthesia (Stockard and Bickford 1975), and in comparison with enflurane, epileptic seizure activity cannot be provoked by stimulation or hypocapnia (Eger *et al.* 1971). Accordingly isoflurane has been used successfully in patients with status epilepticus where halothane did not eliminate the seizure activity (Ropper *et al.* 1986), and prolonged low flow isoflurane anaesthesia has been proposed for the control and treatment for status epilepticus (Kofke *et al.* 1985). Although pre- and postoperative seizure activity has been associated with the administration of isoflurane (Hymes 1985, Harrison 1986), isoflurane as the offending agent has been questioned (Keats 1985). During recovery from isoflurane anaesthesia an EEG arousal phenomenon occurring immediately before awakening has been observed. This phenomenon does not occur during enflurane and seldom during halothane anaesthesia (Erdmann and Brandt 1986). Supplementation with nitrous oxide during 1 MAC isoflurane, enflurane and halothane anaesthesia induces marked EEG changes. Generally, a shift to lower frequencies was accompanied by a pronounced reduction of amplitude resulting in a decrease of total EEG power (Brandt *et al.* 1985).

During isoflurane anaesthesia, a dose-dependent increase in evoked potential latencies, a decrease in amplitudes and a prolongation of central conduction time has been observed (Peterson *et al.* 1984, Sebel *et al.* 1986, Heneghan *et al.* 1987).

In patients undergoing isoflurane anaesthesia for carotid endarterectomy, critical CBF, defined as flow below which the majority of patients developed ipselateral EEG changes of ischaemia within 3 min of carotid occlusion was found to be 8–10 ml/100 g/min (Messick et al. 1987). This value is considerably lower compared with the values obtained during halothane anaesthesia (Sharbrough et al. 1973). In a retrospective study from the same group, including 2000 patients subjected to carotid endarterectomy in either halothane, enflurane or isoflurane anaesthesia, the values for the ischaemic threshold for isoflurane and halothane were confirmed, and for enflurane the ischaemic threshold was found to be 15 ml/100 g/min. The incidence of ischaemic EEG changes was significantly less during isoflurane anaesthesia (18%), than during enflurane (26%) or halothane (25%) (Michenfelder et al. 1987). These studies, together with the experimental studies of the effect of isoflurane on cerebral circulation and metabolism, and studies concerning the central hemodynamics showing stabilization of cardiac output, a dose-dependent decrease in MABP and peripheral vasodilation has increased interest in isoflurane as an agent used for controlled hypotension in neuroanaesthesiological practice. Accordingly clinical studies indicate that isoflurane is a safe and effective agent for the induction of hypotension in neurosurgical patients (Campkin and Flinn 1986).

In surgery for cerebral aneurysm in isoflurane-induced hypotension to MABP 50–60 mm Hg, cardiac output is unchanged (Lam and Gelb 1983), and the pulmonary shunt fraction is unaffected (Nicholas and Lam 1984). Recently, three studies of CBF and metabolism during isoflurane-induced hypotension in patients subjected to craniotomy for cerebral aneurysm have been published (Newman et al. 1986, Roth et al. 1986, Madsen et al. 1987c). In all three studies, CBF was unchanged during isoflurane-induced hypotension, while $CMRO_2$ was reduced by 9–58%, and CBF was considerably higher, compared with the threshold values of cerebral ischaemia of 10 ml/100 g/min obtained during carotid endarterectomy (Messick et al. 1987). In the study by Madsen et al. (1987c) a significant increase in CBF during the posthypotensive period, as compared with the prehypotensive period was observed. This increase was not associated with changes in $PaCO_2$ or MABP, and it was suggested that increased isoflurane concentration was still present in cerebral tissue at the time of investigation, but hyperaemia, secondary to regional cerebral ischaemia cannot be excluded, although global CBF was far above ischaemic threshold.

The effects of fast and slow increase in isoflurane concentration on cerebral haemodynamic has recently been studied during isoflurane-induced hypotension to MABP 60 mm Hg by repeated determination of $AVDO_2$ in patients undergoing cerebral aneurysm surgery. The results indicate that rapid induction, in comparison with slow induction, is accompanied by an early and more prolonged decrease in $AVDO_2$, suggesting a state of relative luxury-perfusion (Haraldsted et al. 1989).

The new volatile agent *Sevoflurane* has a very low blood/gas solubility ratio compared to isoflurane (0.59 versus 1.41), suggesting that it may offer some advantages over isoflurane in neuroanaesthesia (Wallin et al. 1975). In studies in rabbits, the effects of sevoflurane on CBF, ICP, $CMRO_2$, and the EEG are similar to those of isoflurane (Scheller et al. 1988).

Conclusion

Isoflurane in comparison with halothane and enflurane, possesses interesting qualities which make it an inhalation agent to be considered in neuro-anaesthetic practice. Presently, experimental and clinical studies indicate that isoflurane induces a moderate dose-dependent increase in CBF, CBV and ICP, and a decrease in CVR and $CMRO_2$. The CO_2 reactivity generally is preserved, and the changes in CBF, CBV and ICP can be attenuated by controlled hyperventilation. Furthermore, the pronounced suppression of $CMRO_2$ suggest a brain protective effect, and the cerebral autoregulation compared to halothane anaesthesia is only partly impaired. In comparison with halothane and enflurane, isoflurane suppresses the EEG dose-dependently, and at concentrations of 2–3% the EEG is isoelectric. The effect of isoflurane on the peripheral resistance can be used therapeutically when controlled hypotension is decided. However, isoflurane may decrease MABP and CPP, especially in hypovolaemia and patients with reduced cardiac reserves, necessitating reduction in inspiratory concentration or infusion of considerable amounts of plasma expanders or crystalloids.

Nitrous Oxide

Repeated studies in mice and rats (Carlsson et al. 1976b, Harp et al. 1976, Ratcheson et al. 1977, Dahlgren et al. 1981, Ingvar and Siesjö 1982), have indicated that CBF and $CMRO_2$ are unchanged during N_2O administration. However, in dogs an increase in both blood flow and oxygen uptake have been found (Theye and Michenfelder 1968a, Sakabe et al. 1978, Oshita et al. 1979, Gibson and Duffy 1981, Fitzpatrick and Gilboe 1982). In dogs, CBF is increased during normocapnic nitrous oxide administration, but not during hypocapnia, and in the same animal nitrous oxide increased ICP both in normal animals and animals subjected to ICP–hypertension by inflation of an epidural balloon, both during normo- and hypocapnia

(Albin *et al.* 1986). In rats studied with the 14-C deoxyglucose method the local cerebral glucose utilization was increased by 15–25% in subcortical structures (red nucleus, thalamus, geniculate bodies and superior colliculus), but decreased in nucleus accumbens and sensorimotor cortex by comparable amounts (Ingvar and Siesjö 1982, Pelligrino *et al.* 1984). *In vitro* studies of cerebral cortex mitochondria from the goat show that the increased oxygen consumption that accompanies nitrous oxide anaesthesia cannot be attributed to a direct effect of nitrous oxide on mitochondrial respiration (Becker *et al.* 1986). In dogs, the concentration of ATP and lactate is unchanged during N_2O anaesthesia (Michenfelder *et al.* 1970).

In mice subjected to hypoxia, nitrous oxide reduces survival and reverses thiopental-induced prolongation of survival time (Hartung and Cottrell 1987). On the other hand Milde (1988) in a similar study found that nitrous oxide had no effect on survival time. As stated by Artru (1988b) methodological errors including ambient temperature and the inspiratory concentration of oxygen might be responsible for the discrepancy in results.

In conclusion, nitrous oxide only influences cerebral circulation and metabolism minimally but a certain degree of activation can occur, if stress factors are not eliminated (Carlsson *et al.* 1976b). These considerations are interesting because nitrous oxide in experimental studies as well as human investigations is used for sedation and it has been observed that CBF and $CMRO_2$ in paralysed animals increase considerably when nitrous oxide is exchanged with nitrogen, while this increase is not observed in rats subjected to adrenectomy and rats undergoing treatment with beta-blocking agents (Carlsson *et al.* 1975a).

In cats where halothane or isoflurane were used as basis anaesthetic, nitrous oxide induces a significant increase in CBF (Drummond and Todd 1985, Hansen *et al.* 1988c), and the increase in CBF elicited by nitrous oxide is highest at high end-tidal concentrations of halothane or isoflurane (Drummond *et al.* 1987). Moreover, studies in rats have demonstrated that the addition of nitrous oxide to background halothane anaesthesia was associated with dilatation of cerebral vasculature and an increase in ICP (Seyde *et al.* 1986). In rabbits the increase in CBF and ICP are accompanied by an increase in EEG frequency and decrease in amplitude and these changes as well as the increase in CBF and ICP were also observed during hypocapnia (Todd 1987, Kaieda *et al.* 1989a). Contrary to these observations, nitrous oxide combined with diazepam in rats elicits an associated and pronounced decrease in CBF and $CMRO_2$ (Carlsson *et al.* 1976b, Carlsson and Chapman 1981). The effects of combinations of nitrous oxide with enflurane or isoflurane on blood flow, metabolism and outcome in experimental ischaemia have been described in the sections of enflurane and isoflurane.

In humans, CBF is unchanged or increased during nitrous oxide administration, while $CMRO_2$ is decreased by 10–20% (Wollman *et al.* 1965, Alexander *et al.* 1968, Smith *et al.* 1970). Recent studies in adult volunteers indicate a dose-dependent increase in global CBF especially in the anterior regions, while a decrease in the posterior cortical regions was observed (Samra *et al.* 1988). Administered together with halothane, nitrous oxide gives rise to an increase in jugular venous oxygen tension, suggesting cerebral vasodilatation (Sakabe *et al.* 1976). During induction of anaesthesia in patients with space-occupying lesions, nitrous oxide induces an increase in ICP (Henriksen and Balslev-Jørgensen 1973, Misfeldt *et al.* 1974, Moss and McDowall 1979), and an increase in CBF and jugular venous oxygen tension (Sakabe *et al.* 1976). The increase in ICP disappears when nitrous oxide is withdrawn, and during hypocapnia (Misfeldt *et al.* 1974, Balslev-Jørgensen and Misfeldt 1975). It is suggested that the increase in ICP during nitrous oxide induction, may partly be caused by an increase in $PaCO_2$, provoked by a decrease in pulmonary minute ventilation owing to absorption of nitrous oxide in the lungs (Cold 1975).

In humans subjected to painful toothpulp electrical shocks, evoked potential latencies and amplitudes are restored after nalaxone administration (Chapman and Benedetti 1979).

In the sitting and the steep head-up position, nitrous oxide should immediately be withdrawn if clinical signs of air-embolism occur and the use of nitrous oxide should be avoided by pneumoencephalography or when pneumoencephalocele is present. The use of nitrous oxide in neuro-anaesthesiological practice has recently been discussed and it has been argued that the use of i.v. infusion of hypnotic and analgetis agents can suitably be an alternative to techniques of anaesthesia with nitrous oxide (Barker 1987).

3. Hypnotic Agents

Barbiturate: Experimental Studies

In dogs and rats a dose-dependent decrease in CBF and $CMRO_2$, associated with an increase in CVR has been observed after bolus injection of thiopentone (Michenfelder 1974, Carlsson *et al.* 1976, Stullken *et al.* 1977), and phenobarbitone (Nilsson and Siesjö 1975). Through the action of barbiturate on CVR and CBV, ICP will decrease. The suppression of $CMRO_2$ by thiopentone injection reach a plateau of 50% of control, when EEG is isoelectric (Michenfelder 1974, Kassell *et al.* 1980, Steen *et al.* 1983). However, in dogs subjected to thiopentone combined with hypothermia to 30 C, $CMRO_2$ decreases to about 70% of control values (Lafferty *et al.* 1978). At an even higher degree of hypothermia (14–18 C), EEG becomes isoelectric and $CMRO_2$ is reduced to 7–14% of control. Under these circumstances $CMRO_2$ is unaffected by barbiturate (Steen *et al.* 1983). During continuous thiopentone infusion in dogs and goats, loss of consciousness occurs when CBF and $CMRO_2$ are reduced by 23–30% and emergence from anaesthesia is observed while $CMRO_2$ is still depressed by about 20% (Stullken *et al.* 1977, Albrecht *et al.* 1977). In dogs subjected to prolonged deep barbiturate anaesthesia for 90 min, until burst suppression on EEG, methohexitone and thiopentone induces the same decrease in CBF and $CMRO_2$; however, when the barbiturate infusions were discontinued, recovery occurred earlier with methohexitone than with thiopentone and $CMRO_2$ and CBF returned more rapidly toward control values (Boardine *et al.* 1984). In dogs, pretreatment with thiopentone immediately before induction of anaesthesia is associated with acute tolerance of $CMRO_2$ in spite of higher concentrations of thiopentone in CSF and blood (Altenburg *et al.* 1969). On the other hand, chronic thiopentone infusion for 24 hours in dogs, is followed by sustained increase in $CMRO_2$ occurring after 3 hours in spite of unchanged blood levels of thiopentone (Gronert *et al.* 1981). In rats, the effect of thiopentone on CBF and $CMRO_2$ is similar in young and older animals (Baughman *et al.* 1986). Following high doses of thiopentone injection in dogs and rats until isoelectric EEG and a decrease in $CMRO_2$ averaging 50%, the concentrations of ATP, ADP and AMP are unchanged, while the concentrations of glycogen and phosphocreatine are increased, indicating an unchanged or increased energy charge level, and that the cerebral metabolic effect of barbiturate are secondary to functional effects (Michenfelder 1974, Carlsson *et al.* 1975b).

Studies of regional glucose utilization in rats indicate that barbiturates induce a universal metabolic depression, partly excluding nucleus interpeduncularis, nucleus habenula and tractus habenulo-interpeduncularis (Herkenhan 1981, Hodes *et al.* 1983). In dogs a shifting of percentage contribution of flow to slow compartment (white matter) was observed after phenobarbitone infusion. It is suggested that this selective shunting of blood to white matter might explain the fall in ICP and protection of deep white matter observed by many authors (Laurent *et al.* 1982). In comparison with the cerebral cortex, the depression of glucose utilization in the spinal cord is much less (Crosby *et al.* 1984). During combined use of nitrous oxide and thiopentone at levels with EEG activity, glucose utilization is higher in many structures, including the midbrain reticular formation, than that observed with thiopentone alone. These changes were not observed with high thiopentone injection with isoelectric EEG and indicate that nitrous oxide acts as a cerebral metabolic stimulant in the presence of cerebral function during thiopentone anaesthesia (Sakabe *et al.* 1985).

Studies in rats indicate an inhibition of cerebral phosphofructokinase activity by barbiturates, that endogenous substrates are mobilized from existing carbohydrate and amino-acid pools and that thiopentone solubilizes bound hexokinase activity thereby inhibiting energy metabolism (Nilsson and Siesjö 1974, Chapman *et al.* 1978, Krieglstein *et al.* 1981).

In baboons subjected to MCAO, the ultra-short acting barbiturate, methohexitone produces an elevation of CBF in regions of cortex where flow is below evoked potential threshold. This flow increase most likely results from a reduction in flow in relatively well-perfused cortical regions as a consequence of a reduction in metabolic rate and vasoconstriction by the barbiturate; blood flow is therefore diverted into relatively ischaemic regions (an *inverse steal*) (Branston *et al.* 1979). In cats subjected to MCAO, pentobarbitone infusion will increase cerebral oxygen availability in poorly perfused cerebral cortex (Feustel *et al.* 1981).

During high dose thiopentone infusion in dogs until burst suppression level in the EEG, a profound degree of vasoconstriction, equivalent to that produced by hypocapnia with $PaCO_2$ 20 mm Hg occurs. Under these circumstances the CO_2 reactivity is fairly low in the $PaCO_2$ level ranging from 30–40 mm Hg, and absent in the range of 20–30 mm Hg, suggesting that hyperventilation to levels of $PaCO_2$ less than 30 mm Hg not effectively increases the degree of vasoconstriction (Kassell *et al.* 1981).

Barbiturate initially produces fast EEG activity of 15–30 Hz, with the largest amplitudes frontally. The dominant frequency of these fast waves.

and the rate at which this dominant frequency is attained both decrease as the total cumulative dose of barbiturate is augmented. The polymorphic delta waves which initially alternate with the fast activity become more constant as the cumulative dosage rises and become associated with 10–15 Hz spindles. Further increase in total dose causes bursts of activity alternating with runs of suppression and at an even higher dose total electric silence occurs (Stockard and Bickford 1975).

During hypoxic studies in mice, barbiturate has been shown to improve survival and decrease mortality (Arnfred and Secher 1962), and *in vitro* studies of slices from hippocampus tissue exposed to periods of hypoxia have shown that addition of thiopentone to the perfusion medium increases the duration of hypoxia that was survived by CA1 pyramidal cells (Aitken and Schiff 1986). The survival time in hypoxic mice is increased by mild hypothermia and decreased by hyperthermia, and when mild hypothermia is combined with pentobarbitone, survival time was generally further increased (Artru and Michenfelder 1981). This protective effect of barbiturate is bound to its anaesthetic effect and seems distinct from the anticonvulsant effect (Steen and Michenfelder 1979). As judged by metabolic criteria by measurements of cerebral tissue ATP, phosphocreatine and lactate in rats subjected to unilateral carotid ligation, hypothermia offers better cerebral protection than does phenobarbitone that gives the same decrease in CBF and $CMRO_2$ (Hägerdal *et al.* 1978).

In cats subjected to ICP hypertension by inflation of an epidural balloon, the increase in ICP during inflation was less in animals pretreated with barbiturate than untreated animals and in the postdeflation period, the untreated animals developed higher ICP increase than pretreated animals, suggesting a protective effect against postcompression brain swelling (Bricolo *et al.* 1981). By following intracellular brain pH in monkeys subjected to MCAO in either halothane or thiopentone anaesthesia, it has been shown that the decrease in pH is less pronounced in barbiturate anaesthetized animals, and after CBF had been restored, brain pH returned toward normal under thiopentone anaesthesia but continued to deteriorate under halothane anaesthesia (Anderson and Sundt 1983).

Several experimental studies of focal and global cerebral ischaemia in cats, dogs and monkeys have shown that pre-treatment with barbiturate offer a brain protective effect by improving survival and diminishing neuropathological changes (Hoff *et al.* 1973, Smith *et al.* 1974, Hankinson *et al.* 1974, Hoff *et al.* 1975, Corkill *et al.* 1976, Safar *et al.* 1976, Michenfelder *et al.* 1976, Corkill *et al.* 1978, Bleyart *et al.* 1978, Todd *et al.* 1982, Gelb *et al.* 1986). In rats subjected to asphyxia and arterial hypotension (Nilsson 1971) and dogs subjected to hypoxia (Michenfelder and Theye 1973), barbiturate preserves the cerebral energy charge and the lactate production is decreased. Furthermore, in gerbils subjected to unilateral carotid ligation for one hour, neuropathological examination in animals allocated to thiopentone injected one hour after removal of the arterial clamp revealed less extensive neuronal ischaemic cell changes (Levy and Brierley 1979). In baboons and dogs subjected to temporary MCAO (Selman et al. 1981, Yonas et al. 1981) or extracranial–intracranial by-pass operations (Spetzler et al. 1982), it has been demonstrated that sufficient reperfusion after focal cerebral ischaemia is of utmost importance in the protection against cerebral ischaemic damage. On the other hand, other experimental studies in dogs and monkeys subjected to complete global ischaemia have stated that barbiturates do not protect the brain (Steen et al. 1979, Snyder et al. 1979, Gisvold et al. 1984, Koch et al. 1984), and under these circumstances barbiturates if at all (Steen et al. 1978) only induce a very moderate delay in the depletion of cerebral energy reserves (Nordström and Siesjö 1978). Thus, experimental data do not at present support the use of barbiturate in the treatment of global ischaemia (Shapiro 1985).

It has been postulated that free radicals are released in ischaemic cerebral tissue, giving rise to irreversible tissue damage (Michenfelder et al. 1976), and that barbiturate should act as a free radical scavanger, thus having its protective effect (Smith et al. 1974, Flamm et al. 1977). However, this hypothesis has recently been rejected (Smith et al. 1980, Siesjö 1984), and it is presently accepted that the protective effect of barbiturate is caused by the suppression of cerebral metabolism (Astrup 1982, Siesjö 1984).

Human Investigations

The experimental data concerning the effect of barbiturates on cerebral circulation and metabolism have been confirmed in clinical studies. Thus, barbiturate induces an associated decrease in CBF and $CMRO_2$ and an increase in CVR (Wechsler et al. 1951, Pierce et al. 1962); and these effects are associated with a decrease in ICP (Søndergård 1961, Gordon 1970).

The EEG changes produced by thiopentone injection have been divided into five patterns by Kiersey et al. (1951). The first pattern is characterized by high amplitudes and fast activity of mixed frequencies varying between 10–30 Hz. The second pattern is a complex of many frequencies but with the presence of predominantly slower waves and much variation in voltages. In the third pattern, progressive suppression of cortical activity is separated by bursts of activity. In the fourth pattern the durations of periods with inactivity is further prolonged, ranging between 3 and 10 seconds. During the fifth pattern, the amplitudes fall below 25 microvolts and long periods with isoelectric EEG are recorded.

Reviews concerning barbiturate treatment in brain ischaemia and high
dose barbiturate therapy in neurosurgery and intensive care have recently
been published (Shapiro 1985, Piatt and Schiff 1984). In the following
clinical studies of barbiturate in the neuroanaesthetic practice and in the
care of neuro-intensive patients will be referred to.

In the neuroanaesthesiological practice, thiopentone is used during
induction of anaesthesia before the use of inhalation agents (Hunter 1972,
Astrup et al. 1984, Madsen et al. 1987a) and before neuroleptanaesthesia
(Arner and Gordon 1976, Cold et al. 1988). In clinical studies of ICP
hypertension, preoperative loads of thiopentone (1.5–3.0 mg/kg) induce
a significant decrease in ICP (Shapiro et al. 1973), but although studies of
thiopentone supplemented anaesthesia for craniotomy have been promising, the postoperative recovery is prolonged (Hunter 1972). Preoperatively,
after opening of the dura, thiopentone in a dose of 4–8 mg/kg during
maintenance of anaesthesia with nitrous oxide, halothane 0.5% gives rise to
a small but significant decrease in $CMRO_2$ by 8%, and it has been
suggested that although small, this decrease in metabolism may have some
protective effect (Astrup et al. 1984).

In uncontrolled studies, peroperative thiopentone loads have been
advocated in surgery for cerebral aneurysm (Hoff et al. 1977, Sokoll et al.
1982, Bendtsen et al. 1984, Belopavlovic et al. 1985), and uncontrolled
studies suggest that the frequency of cerebral spasms in patients with
cerebral aneurysms might be reduced after deep barbiturate anaesthesia
(Belopavlovic et al. 1985). Studies of central hemodynamics during
aneurysm surgery with a thiopentone dose of 20–34 mg/kg indicate a reduction of cardiac output and MABP, without generalized evidence of
compromised tissue perfusion (Sokoll et al. 1982, Bendtsen et al. 1983,
Todd et al. 1985), and similar conclusions have been drawn during
methohexitone anaesthesia (Todd et al. 1984a) and pentobarbitone
anaesthesia (Traeger et al. 1983, Todd et al. 1987).

In patients subjected to carotid endarterectomy (Moffat et al. 1983,
Hicks et al. 1986) and extracranial–intracranial by-pass operations (Lawner
and Simeone 1979, Lobe and Brauer 1983), barbiturate anaesthesia has
also been advocated and it has been argued that barbiturate might protect
the brain.

In patients with severe head injury with persistent ICP hypertension in
spite of hyperventilation, thiopentone might induce a rapid decrease in ICP
accompanied by an increase in CPP and sustained ICP reduction could be
maintained for several days by combining thiopentone therapy and hypothermia to 30 C (Shapiro et al. 1974). Other clinical studies support these
findings (Sidi et al. 1983). In several clinical studies of patients with severe
head injury, barbiturate sedation together with hyperventilation has been

advocated in the control of ICP hypertension, and this regime has been claimed to improve outcome (Saul and Ducker 1982). While uncontrolled (Rockoff et al. 1979), and controlled studies (Eisenberg et al. 1988) seem to support this view other uncontrolled studies (Yano et al. 1986), and controlled studies (Ward et al. 1985) did not confirm this finding.

In a controlled study of thiopentone treatment after cardiac arrest no improvement of outcome was disclosed (Abramson et al. 1983).

In patients subjected to a cardiopulmonary by-pass operation for open ventricle operation a preliminary controlled study indicated that thiopentone might reduce postoperative neuropsychiatric dysfunction, supposed to be caused by microembolization (Nussmeier et al. 1986). This observation has given rise to optimism and it has been proposed that thiopentone treatment during open heart surgery should not be denied the patients (Michenfelder 1986). However, the issue has recently been debated (Scheller et al. 1986), and it has been suggested that other factors like hypocapnia might decrease the risks of air embolization, owing to the general reduction in CBF during a by-pass operation (Henriksen 1986). Furthermore, in a recent study by Nussmeier and Fish (1987), the changes in neuropsychiatric function after a cardiopulmonary by-pass operation were not mitigated by thiopentone or prostaglandin (PGI-2) injection. Only the age of the patients was found to be a predictor of postoperative decrements in performance.

Surgery for arterio-venous malformation can pre- or postoperatively be complicated by massive multifocal bleeding and/or edema. This phenomenon called normal perfusion pressure breakthrough has experimentally been studied by Spetzler et al. (1978), who found that diversion of blood from the arterio-venous malformation into adjacent, maximally dilated and non-autoregulated small vessels was the pathophysiological reason. The occurrence of this complication is catastrophic but preliminary studies indicate that deep thiopentone anaesthesia with loading doses of 10–20 mg/kg can prevent this complication and reduce the formation of edema (Day et al. 1982, Marshall and Sang U 1983).

Methohexitone a barbiturate derivate acts like thiopentone and gives rise to an associated decrease in CBF and $CMRO_2$. In rabbits subjected to global ischaemia, pre-treatment with methohexitone at the onset of isoelectric EEG provides protection from ischaemic damage and the animals recovered dramatically (Yatsu et al. 1972) and in baboons subjected to MACO, methohexitone induces an inverse steal phenomenon (Branston et al. 1979). On the other hand, methohexitone gives rise to electroencephalographic activation in patients with psychomotor epilepsia (Musella et al. 1971), and enhances electroencephalographic delineation of epileptogenic foci during surgery for uncontrollable epilepsia (Ford et al. 1982). In a clinical study by Todd *et al.* (1984a) where continuous infusion of methohexitone was used for patients undergoing operation for arteriovenous malformation, postoperative seizures occurred in 3 out of 11 patients. Although methohexitone is a chemical convulsant (Musella *et al.* 1971), this finding is suggested to be caused by acute methohexitone withdrawal.

Human studies indicate, that after intravenous bolus injection of methohexitone the maximal reduction of CBF occurs within 30 sec, and CBF is normalized within 10 min (Herrschaft *et al.* 1975).

Althesin

This hypnotic agent consisting of a mixture of alphaxalone and alphadolone acetate dissolved in an emulsifying agent (Cremophor EL) is presently not in use in the scandinavian countries because of the relative high frequency of anaphylactoid reaction but the effects of the drug on cerebral circulation and metabolism have been carefully studied and shall therefore be cited.

Early *animal experiments* have shown that a bolus injection of althesin gives rise to an associated decrease in CBF and $CMRO_2$ and a reduction in ICP (Pickerodt *et al.* 1972, Takahashi *et al.* 1973, Fitch *et al.* 1978, Keaney *et al.* 1978). These changes are followed by suppression of the EEG (Prior *et al.* 1978). Comparative studies of pentobarbitone and althesin anaesthesia in cats suggest that the two drugs depress neocortical function in a similar manner but in the hippocampus the EEG spikes were less prominent during pentobarbitone than during althesin anaesthesia (Kayama 1974). The decrease in CBF, effected by a bolus injection occurs after 30 sec and subsides after 10–15 min (Pickerodt *et al.* 1972). Other experimental studies in baboons have shown that the decrease in CBF occurs 2 sec after slowing of the EEG, while cortical extracellular fluid pH began to increase about 10 sec after the EEG changes, indicating that the change in extracellular pH cannot have initiated the decrease in blood flow which follows the suppression in cerebral metabolism (McDowall *et al.* 1979). The CO_2 reactivity is present during althesin anaesthesia (Fitch *et al.* 1978). In studies of regional cerebral glucose utilization in rats regional differences have been found. Thus the glucose utilization was reduced considerably in the forebrain (cerebral cortex), while the decrease in the hindbrain was much less. These observations suggest that naturally occurring alphaxalone receptors might be present in the brain (Davis *et al.* 1984).

Human studies have confirmed the findings of animal experiments. Thus, a dose-dependent decrease in CBF and $CMRO_2$ occur (Renou *et al.*

1976, Sari *et al.* 1976, Rasmussen *et al.* 1978), associated with a decrease in ICP (Turner *et al.* 1973). Simultaneously with the suppression of cerebral oxygen uptake, the EEG is changed, first showing low frequency activity, later on burst suppression activity, and finally isoelectric recording (Prior *et al.* 1978, Henderson *et al.* 1982).

Accordingly, althesin infusion has been proposed in the treatment of status epilepticus (Munari *et al.* 1979). In patients with severe head injury an inverse CBF increase has been demonstrated after bolus injection of althesin (Rasmussen *et al.* 1978) and heavy althesin sedation has been shown to be more effective in controlling ICP hypertension compared with barbiturates, although the clinical outcome is not improved (Bullock *et al.* 1986). Recent studies in severe head injury suggest that althesin bolus injection is especially effective in reducing ICP and increase CPP, when EEG cortical activity is present (Procaccio *et al.* 1988). During craniotomy for cerebral tumors, althesin induces a dose-dependent decrease in $CMRO_2$

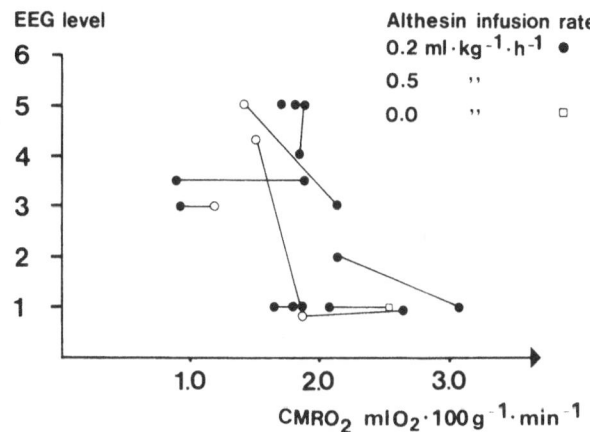

Fig. 3. Correlation between $CMRO_2$ and EEG level in 10 paired studies in 10 patients. The studies were performed during craniotomy for supratentorial cerebral tumors in 10 patients anaesthetized with continuous althesin infusion, nitrous oxide 67% and fentanyl. The EEG levels were classified as follows: Level *1*: Continuous background activity of fairly constant voltage with any combinations of frequencies, but without periods of either partial or total suppression. Level *2*: Periods of less than 1 s total or subtotal suppression, separated by bursts of activity usually of 100–300 μV. Level *3*: Periods of 1–3 s duration of total, occasionally subtotal suppression, separated by bursts of activity of 100–300 μV. Level *4*: Periods of at least 3 s total suppression separated by bursts of activity usually of 50–100 μV. Level *5*: Periods of at least 3 s total suppression, separated by very brief bursts of low voltage less than 50 μV. Level *6*: No evidence of any cerebral activity even with high gain (Br J Anaesth 1985: 57: 369–374, with permission)

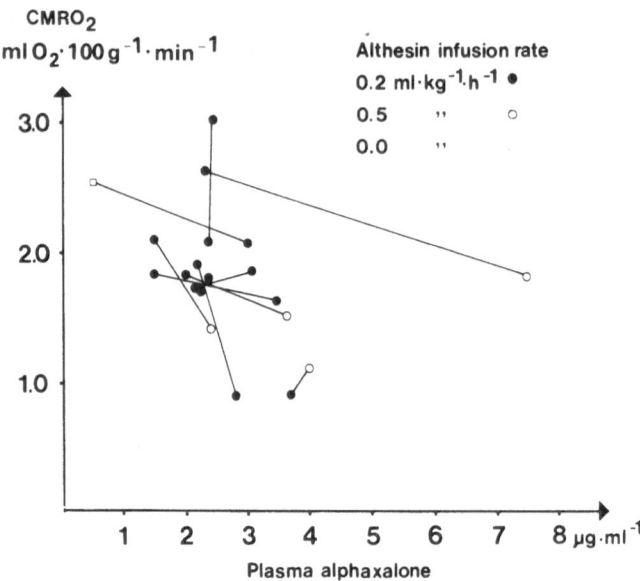

Fig. 4. Correlation between Plasma Alphaxalone (µg/ml) and CMRO₂ (ml O₂/100 g/min) in 10 paired studies performed during craniotomy in patients with supratentorial cerebral tumors anaesthetized with continuous althesin infusion, nitrous oxide 67% and fentanyl (Br J Anaesth 1985: 57: 369–374, with permission)

and suppression in EEG, correlated to the concentration of plasma al-faxolone. However, great inter- and intraindividual differences in response were observed making conclusion concerning the level of oxygen uptake from the EEG recordings or plasma alphaxalone impossible (Bendtsen *et al.* 1985). The correlations between CMRO₂ and the EEG levels and the correlations between plasma alphaxalone and CMRO₂ are shown in Figs. 3 and 4.

Etomidate

In *experimental studies* etomidate like barbiturate and althesin decreases CBF and CMRO₂ and suppresses the EEG. In dogs the EEG changes elicited by etomidate are similar to barbiturate (Wauquier *et al.* 1978). During continuous infusion of etomidate in dogs, CMRO₂ decreases until there is cessation of neuronal function as reflected by the onset of isoelectric EEG. This occurred when CMRO₂ was 48% of control. In the same study CBF initially decreased precipitously and independently of the changes in

$CMRO_2$ (Milde *et al.* 1985). In rats, etomidate has a marked effect on glucose utilization of the brain. Like althesin this effect is most pronounced in the forebrain, while the hindbrain is minimally affected, suggesting that the anaesthetic effect of these drugs might be mediated by receptors. Furthermore, the decrease in glucose utilization was not dose-dependent, thus 1 and 12 mg/kg depressed the glucose utilization equally (Davis *et al.* 1986).

During hypoxia and global ischaemia in mice and dogs, etomidate has a brain protective effect (Wauquier *et al.* 1981, Wauquier 1982), and in neurohistological studies in rats subjected to lethal cyanide doses and in models of global ischaemia it has been shown that animals in pre-treatment with etomidate preserve normal brain morphology (Ashton *et al.* 1981, Van Reempts *et al.* 1982). In dogs subjected to incomplete cerebral ischaemia, the concentrations of ATP and phosphocreatine are maintained at higher levels than in untreated dogs, suggesting that etomidate improves tolerance of the brain to ischaemia (Milde and Milde 1986). Maintenance of high energy metabolites has also been demonstrated in experiments of severe hypoxia in rats (Keykhah *et al.* 1986). However, in rats subjected to incomplete ischaemia, high dose in contrast to low dose etomidate provokes spiking EEG activity without further depression of $CMRO_2$, and a worsening of outcome following cerebral ischaemia (Baugmann *et al.* 1989a).

In *Human studies* etomidate induces an associated decrease in CBF and $CMRO_2$ (Herrschaft *et al.* 1975, Van Aken and Rolly 1976, Renou *et al.* 1978), and a decrease in ICP (Cunitz *et al.* 1973, Schulte an Esch *et al.* 1978, Moss *et al.* 1979).

In spite of involuntary movements during etomidate induction, a dose-dependent suppression of the EEG activity occurs (Ghomein and Yamada 1977, Doenicke *et al.* 1982) and at deep etomidate anaesthesia the EEG is isoelectric. However, seizure activity has been observed during etomidate anaesthesia (Krieger *et al.* 1985) and prolonged etomidate sedation (Grant and Hutchison 1983) and in patients undergoing craniotomy for surgical removal of their seizure focus (Gancher *et al.* 1984) indicating a selective activation of the cerebral cortex by etomidate. Consequently some authors suggest that etomidate should be avoided in patients with epilepsia (Ebrahim *et al.* 1986).

During induction of anaesthesia with etomidate in patients with intracranial space-occupying lesions, ICP is well controlled (Belopavlovic and Buchthal 1983), and continuous infusion of etomidate has been advocated for neuroanaesthesia (Sear *et al.* 1984, Batjar *et al.* 1988). During continuous etomidate infusion, monitoring of EEG indicates an increased sensitivity to etomidate in the elderly patients (Arden *et al.* 1986). In patients

subjected to craniotomy for small supratentorial cerebral tumors, continuous etomidate infusion supplemented with nitrous oxide and fentanyl gives
rise to a dose dependent decrease in CBF and $CMRO_2$, and the relative
CO_2 reactivity is preserved (Cold *et al.* 1985). In most patients a fairly good
correlation between the changes in $CMRO_2$, the EEG and the plasma
concentration of etomidate were found. However, great inter- and intraindividual differences were observed, making statements concerning the
suppression of cerebral oxygen uptake from the EEG recordings impossible
(Cold *et al.* 1986). The correlations between $CMRO_2$ and EEG levels,
between plasma etomidate and EEG levels and between plasma etomidate
and $CMRO_2$ are shown in Figs. 5–7.

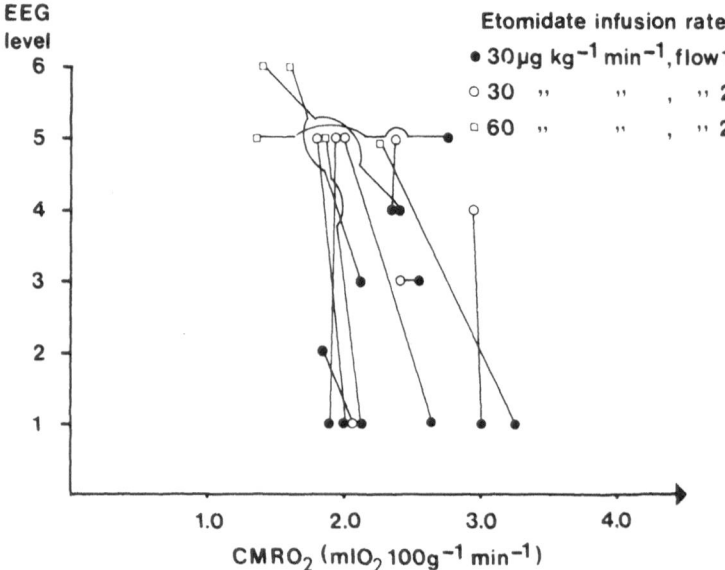

Fig. 5. Correlation between $CMRO_2$ (ml O_2/100 g/min) and EEG levels in 14
paired studies performed in patients subjected to craniotomy for supratentorial
cerebral tumors. The patients were anaesthetized with continuous etomidate
infusion, supplemented with nitrous oxide 67% and fentanyl. The EEG levels were
classified as follows: Level *1*: Continuous background activity of fairly constant
voltage with any combination of frequency, but without periods of either partial or
total suppression. Level *2*: Periods of less than 1 s of total of subtotal suppression,
separated by bursts of activity usually of 100–300 μV. Level *3*: Periods of 1–3 s
duration of total, occasionally subtotal suppression, separated by bursts of activity
of 100–300 μV. Level *4*: Periods of at least 3 s total suppression, separated by
bursts of activity of 50–100 μV. Level *5*: Periods of at least 3 s total suppression
separated by very brief bursts of low voltage less than 50 μV. Level *6*: No evidence
of any cerebral activity even with high gain (Acta Anaesthesiol Scand 1986: 30:
159–163, with permission)

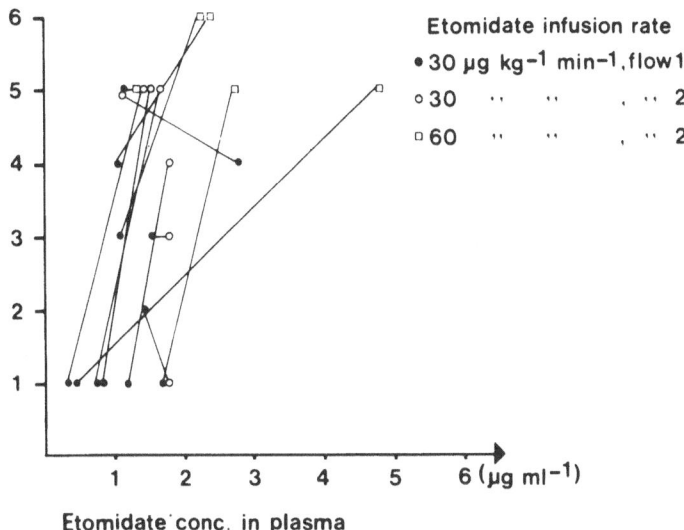

Fig. 6. Correlation between etomidate concentration in plasma (μg/ml), and the EEG level in 14 paired studies performed in patients subjected to craniotomy for supratentorial cerebral tumors. The patients were anaesthetized with continuous etomidate infusion, supplemented with nitrous oxide 67% and fentanyl. The EEG levels were classified as follows: Level *1*: Continuous background activity of fairly constant voltage with any combinations of frequency, but without periods of either partial or total suppression. Level *2*: Periods of less than 1 s of total or subtotal suppression, separated by bursts of activity usually of 100–300 μV. Level *3*: Periods of 1–3 s duration of total, occasionally subtotal suppression, separated by bursts of activity of 100–300 μV. Level *4*: Periods of at least 3 s total suppression, separated by bursts of activity of 50–100 μV. Level *5*: Periods of at least 3 s total suppression separated by very brief bursts of low voltage less than 50 μV. Level *6:* No evidence of any cerebral activity even with high gain (Acta Anaesthesiol Scand 1986: 30: 159–163, with permission)

In patients with severe head injury, continuous infusion of etomidate in doses of 5–25 μg/kg/min, partially seems to control ICP hypertension, and additional doses of 0.2 mg/kg always reduce acutely elevated ICP (Prior *et al.* 1983). However, the effect on ICP hypertension is only observed in patients with spontaneous EEG activity (Bingham *et al.* 1985). This effect has also been observed during alfaxalone sedation, and in a controlled study, the two drugs appeared equipotent in decreasing ICP, and preserving the CPP (Dearden and McDowall 1985).

Early experimental studies showed that etomidate suppresses adreno-cortical response to surgical stress in rats (Preziosi and Vacca 1982), and

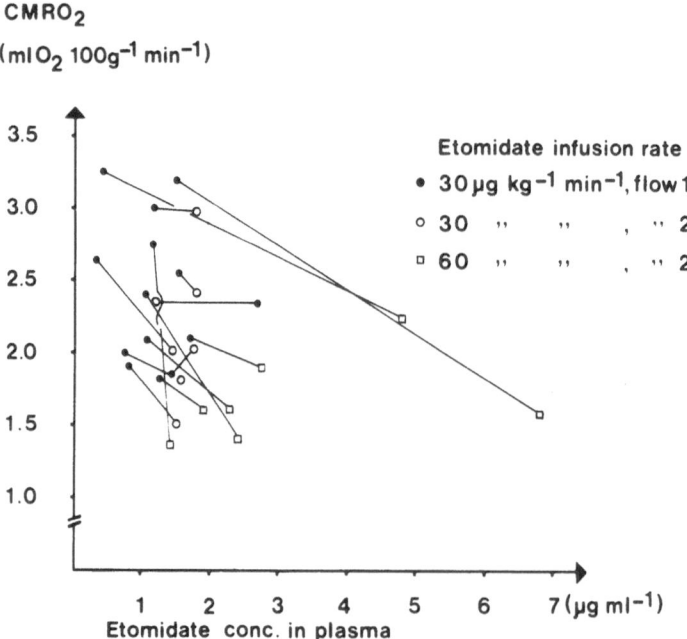

Fig. 7. Correlation between etomidate concentration in plasma (μg/ml) and CMRO$_2$ (ml O$_2$/100 g/min) in 14 paired studies performed during craniotomy in patients with supratentorial cerebral tumors subjected to anaesthesia with continuous etomidate infusion, supplemented with nitrous oxide 67% and fentanyl (Acta Anaesthesiol Scand 1986: 30: 159–163, with permission)

low concentrations of cortisol have been observed in the postoperative period (Finlay and McKee 1982). In studies of postoperative ACTH stimulation, the cortisol and aldosterone responses are decreased in patients subjected to etomidate anaesthesia (Klausen *et al.* 1983, Wagner and White 1984, Fragen *et al.* 1984). Etomidate inhibits adrenal steroidogenesis by producing a concentration-dependent block of the cholesterol side-chain cleavage enzyme and 11-beta-hydroxylase (Wagner *et al.* 1984). In preliminary studies by Ledingham and Watt (1983) the duration of the adrenocortical suppression after a bolus injection of etomidate is about 8–10 h. The suppressive effect of etomidate in patients on treatment with corticosteroids has not been studied. The above-mentioned effects of etomidate on the adrenocortical axis have prompted discussion concerning the use of etomidate in clinical practice. On one hand the attitude is advanced that there is no strong evidence against the continued availability of etomidate for use as an induction agent (Owen and Spence 1984). On the other hand it has been argued that etomidate should be restricted only to those situations where it offers a clinical advantage over other available drugs (Wagner and

White 1984), or should be omitted totally (Longnecker 1984). In comparison with thiopentone, the use of etomidate has two striking advantages; the stability of the central haemodynamics (Criado *et al.* 1980), and the minimal histamine release (Doenicke *et al.* 1973).

Ketamine

In *experimental studies*, ketamine is a cerebral vasodilator, which induces an increase in CBF, $CMRO_2$ and CSF pressure. In dogs ketamine increases CBV concomitant with increases in CSF pressure (Artru and Katz 1989), and in rats the autoregulation is abolished by ketamine (Hickey *et al.* 1989). In rabbits the cerebral vasodilatation is blocked by scopolamine, a cholinergic antagonist, and the increase in CBF after ketamine is additive to the cerebral vasodilator actions of CO_2 and of physiostigmin, suggesting that ketamine activates a cholinergic cerebral vasodilator system (Reicher *et al.* 1987). The metabolic effect of ketamine is influenced by pH in arterial blood; thus experiments in rabbits indicate that the increase in $CMRO_2$ is more pronounced in acidotic animals and the increase in oxygen uptake under these circumstances might enhance the extent of vasodilatation (Oren *et al.* 1987). The effect of ketamine on cerebral circulation is influenced by the anaesthetic procedure applied before ketamine injection. Thus, in paralysed, N_2O anaesthetized goats, CBF and $CMRO_2$ is unchanged (Schwedler *et al.* 1982). In the cat, ketamine induces epileptic EEG of continuous hypersynchronous activity (Kayama and Iwama 1972), and in the same animal, painful stimulation during ketamine, unlike barbiturate, is followed by increased EEG activity in the reticular formation and the limbic system (Tamasy *et al.* 1975). Studies of cerebral glucose utilization in rats have shown a pronounced increase by 70% in the limbic system, especially in the hippocampus, a less pronounced increase by 20–40% in the extrapyramidal system and a decreased metabolism in the somatosensory and auditory systems, with the greatest reduction in the inferior colliculus (40%) (Nelson *et al.* 1980, Crosby *et al.* 1982, Cavazzuti *et al.* 1987). During withdrawal from a bolus injection a stimulation of the glucose utilization has been observed throughout the brain with regional differences ranging between 21 and 78% (Davis *et al.* 1988). Pre-treatment with diazepam decreases the glucose uptake by 20% in the hippocampus, an observation which corroborates with the epileptogenic character of ketamine. In the same study droperidol increased the glucose utilization, especially in the lateral habenula (Oguchi *et al.* 1982). In dogs with normal CSF pressure and no cerebrovascular damage, pre- or posttreatment with diazepam reverses ketamine-induced increase of CBV and CSF pressure (Artru and Katz 1989).

Lightfoote and coworkers (1977) found that cerebral infarction after unilateral carotid ligation in gerbils subjected to ketamine anaesthesia exceeded that occurring with barbiturates; and Albin *et al.* (1989) found that the survival time in rats subjected to hypoxia was shortened. On the other hand studies in rats subjected to near-complete-forebrain ischaemia indicate that ketamine administration over several hours provide significant protection in the hippocampal Ca1 region. Ketamine acts as an antagonist to N-methyl-D aspartate receptors, the activation of which has been found to be implicated in cerebral ischaemic damage. It is suggested that the brain-protective effect of ketamine under these circumstances is mediated through this effect (Church *et al.* 1988). The metabolic effects of ketamine together with the effects on EEG, CVR, CBF and ICP lead to the conclusion that ketamine is considered unsuitable in surgery for space-occupying intracranial lesions, the only exception being the use of ketamine in small doses for induction of anaesthesia in the emergency of haemorrhagic shock (Klose *et al.* 1982).

Ketamine was initially introduced as an anaesthetic agent for neuro-radiological procedures (Corssen *et al.* 1969b): However, *studies in man* showed that ketamine is a cerebral vasodilator. Thus a bolus dosis of 2–3 mg/kg induces an increase in CBF by 50–70% and an increase in $CMRO_2$ by 10–15% (Takeshita *et al.* 1972, Hougaard *et al.* 1974) and an increase in ICP has repeatedly been observed (Evans *et al.* 1971, Gardner *et al.* 1971, List *et al.* 1972, Gibbs 1972, Shapiro *et al.* 1972b). The increase in ICP is also observed in intubated and artificially ventilated patients (Schulte am Esch *et al.* 1978), and in infants with open fontanelles (Crumrine *et al.* 1975). To some extent barbiturates inhibit the increase in ICP (Wyte *et al.* 1972, Belopavlovic and Buchthal 1982). In humans, ketamine anaesthesia is followed by increased EEG activity (Corssen *et al.* 1969a, Thompson 1972, Bennett *et al.* 1973), and in patients with implanted cortical, limbic and thalamic electrodes, ketamine bolus injection provoke seizure activity in the limbic and thalamic areas but not always from the cortical electrodes (Ferrer-Allado *et al.* 1973). It has been suggested that ketamine is not a direct cerebral vasodilator but the increase in CBF is secondary to the increase in neuronal activity and the increase in $CMRO_2$ (Hougaard *et al.* 1974).

As a consequence of the cerebral activation ketamine should be avoided in patients with epilepsia (Winters 1972), and the increase in ICP, CBF and $CMRO_2$, indicate that ketamine should be avoided in neuroanaesthetic practice in all patients with decreased intracranial compliance or intracranial hypertension, the only exception being the use of ketamine in small doses for induction of anaesthesia in the emergency of hemorrhagic shock (Klose *et al.* 1982).

Diazepam

Pharmacological studies have shown that benzodiazepines potentiate the action of gamma-amino butyric acid (GABA), which is an inhibitory neurotransmitter and exerts its actions at specific binding sites (Braestrup and Squires 1977, Möhler and Okada 1977). Benzodiazepine receptors appear to interact closely with certain kinds of GABA receptors (Choi *et al.* 1977). In rats subjected to transient ischaemia an increase in GABA binding receptors in the caudate-putamen has been disclosed (Francis and Pulsinelli 1983). However in gerbils subjected to forebrain ischaemia a marked difference in the distribution between GABA and benzodiazepine binding sites has been found (Onodera *et al.* 1987).

In rats subjected to heavy diazepam sedation a 20–30% decrease in CBF has been observed, while $CMRO_2$ is unchanged; in comparison, supplement of nitrous oxide gives rise to a 50% reduction of both CBF and $CMRO_2$ (Carlsson *et al.* 1976b). In similar animal experiments the energy charge of the cerebrum and the concentrations of ATP and phospho-creatine were unchanged (Carlsson and Chapman 1981). In dogs subjected to lidocaine induced seizures, diazepam reduces CBF and $CMRO_2$ by 50–60% (Maekawa *et al.* 1974), and in rats subjected to ketamine anaesthesia, pre-treatment with diazepam reduces the glucose utilization in the hippocampus (Oguchi *et al.* 1982). In normal dogs, ICP is unchanged after diazepam (Campan and Lazorthes 1976). Experiments in rats suggest that diazepam does not protect the brain from hypoxic insult (Berntman *et al.* 1979).

Human studies of the effects of diazepam on CBF and metabolism are few. In comatose patients with severe head injury, diazepam 15 mg intra-venously, decreases CBF from 45 to 34 ml/100 g/min, and $CMRO_2$ from 1.6 to 1.2 m l O_2/100 g/min (Cotev and Shalit 1975). In patients without organic brain disease, anaesthesia with fentanyl 10 μg/kg, diazepam 10 mg and nitrous oxide 67% decreased CBF and $CMRO_2$ by about 35%; under these circumstances the CO_2 reactivity is preserved (Vernhiet *et al.* 1978). Recently similar results have been obtained in patients subjected to a cor-onary by-pass operation in high-dose fentanyl anaesthesia supplemented with diazepam 0.5 mg/kg (Murkin and Farrar 1989).

Midazolam

In rats midazolam produces a 40–45% decrease in CBF and a 55% decrease in $CMRO_2$; however, when the rats were ventilated with nitrous oxide/oxygen instead of nitrogen oxygen, $CMRO_2$ only decreased by 35%, suggesting that nitrous oxide slightly stimulates cerebral metabolism during midazolam anaesthesia (Hoffman *et al.* 1986). Using the venous outflow technique, similar results have been obtained in dogs (Fleischer *et al.* 1988). In a hypoxic mouse model, midazolam provides greater cerebral protection from hypoxia than diazepam but less protection compared with barbiturates (Nugent *et al.* 1982). In rats subjected to ischaemia by carotid arterial occlusion and hypotension to 30 mm Hg, the mortality rate and neurologic deficit score were identical in midazolam and methohexitone pre-treated animals and in the same study supplements of nitrous oxide increased the mortality rate and neurologic deficit score (Baughman *et al.* 1987b). In the same model high dose midazolam in contrast to etomidate provides better cerebral protection (Baughman *et al.* 1989a). Compared with young rats the cerebral metabolic depression effected by midazolam is more pronounced in aged rats (Baughman *et al.* 1987a). In dogs and goats, subjected to midazolam anaesthesia the benzodiazepine antagonist flumazenil converts the decrease in CBF, $CMRO_2$, CPP and ICP (Fleischer *et al.* 1988, Kochs *et al.* 1988); and these antagonistic effects are dose-dependent (Artru 1989). In comparative studies of the rate of CSF formation and resistance to reabsorption of CSF following the administration of thiopentone, midazolam and etomidate in dogs the rate of CSF formation decreased at high doses of all three drugs; however, the resistance to reabsorption was elevated at high doses of midazolam but decreased at high doses of etomidate and thiopentone (Artru 1988a).

Human studies confirm the results of the animal experiments. Thus, a dose-dependent decrease in CBF and $CMRO_2$ has been disclosed (Larsen *et al.* 1981, Hilfiker and Kettler 1981, Forster *et al.* 1982). In patients with severe head injury, midazolam given as continuous infusion will reduce ICP (Cottrell *et al.* 1982). Injection of the benzodiazepine antagonist flumazenil during midazolam anaesthesia gives rise to a CBF increase (Forster *et al.* 1987) and might provoke an increase in ICP (Chiolero *et al.* 1986a). In studies of continuous midazolam anaesthesia supplemented with fentanyl and nitrous oxide for patients subjected to craniotomy for supratentorial cerebral tumors, a dose-dependent decrease in CBF and $CMRO_2$ was not found; and in the same study flumazenil did not increase CBF or $CMRO_2$ (Knudsen *et al.* 1989). Other studies indicate that injection of flumazenil at the end of anaesthesia is associated with a rapid recovery (Chiolero *et al.* 1986b).

Propofol

In *human studies* propofol given as a bolus injection suppresses CBF, $CMRO_2$ and the cerebral uptake of amino acids (Stephan *et al.* 1987,

Stephan *et al.* 1988). In the same study the CO_2 reactivity was preserved (Stephan *et al.* 1987). Similar changes in CBF and $CMRO_2$ have been observed during continuous infusion with propofol (Vandesteene *et al.* 1988). In patients subjected to craniotomy for supratentorial lesions and given anaesthesia with continuous propofol infusion supplemented with fentanyl, a dose-dependent decrease in $CMRO_2$ was found, while CBF was unchanged (Madsen *et al.* 1989). In comparison with thiopentone and etomidate a significant decrease in arterial pressure follows intubation with propofol anaesthesia (Harris *et al.* 1988). In patients with space-occupying intracranial lesions, propofol induction (1 mg/kg) did not cause any ICP increase and CPP was unchanged, indicating that propofol can be used safely as an induction agent (Hartung 1987, Ravussin *et al.* 1988a). In other studies a bolus dosis of 2.5 mg/kg produced a significant decrease of CPP due to a marked decrease in MABP (Van Hemelrijck *et al.* 1988). In studies of patients for intracranial surgery, propofol induction in comparison with thiopentone–isoflurane induction was accompanied by less MABP reaction to intubation and application of pin head-holder (Ravussin *et al.* 1988b). Combined with alfentanil, total intravenous anaesthesia with propofol in patients with brain tumors provides stable hemodynamic, excellent operating conditions and rapid recovery (Merckx *et al.* 1988). However, in patients with severe head injury, propofol 1 mg/kg will decrease ICP, but MABP and CPP will decrease as well, making the use of propofol in patients with head injury controversial (Hartung 1987). Other studies in severe head injury indicate a coupled reduction by about 30% in CBF and cerebral metabolism and a significant decrease in CPP by 26% (Pinaud *et al.* 1988).

Droperidol

Using the venous outflow technique in dogs, Michenfelder and Theye (1971) found that droperidol in doses of 0.3 mg/kg produced a decrease in CBF by 40%, primarily due to a 30–40% decrease in CVR, and that this effect persisted for the period of observation (60 min). No significant change in $CMRO_2$ was found. In the same study droperidol combined with fentanyl (0.006 mg/kg) resulted in a 23% decrease in $CMRO_2$, CBF was decreased by 50% and CVR increased by 85%. During studies of CBF in dogs subjected to neurolept anaesthesia a decrease of $CMRO_2$ from 5.2 ml to 3.0 ml $O_2/100$ g/min has been observed (Kreuscher 1967). In spontaneously breathing rabbits droperidol combined with fentanyl gives rise to an increase in CBF. However, the change in CBF is correlated to the increase in $PaCO_2$ caused by respiratory depression (deValois and Peperkamp

1971). In experiments in dogs subjected to cryo lesion, the water content in cerebral tissue close to the lesion is low during neurolept anaesthesia. The same low water content was observed during barbiturate anaesthesia, while the content was considerably higher during halothane, enflurane or iso-flurane anaesthesia (Smith and Marque 1976).

Human studies of thalamonal (droperidol and fentanyl) in sedative doses during normocapnia indicate unchanged CBF and $CMRO_2$ (Barker *et al.* 1968, Sari *et al.* 1972), and in patients subjected to controlled ventilation, the combination of droperidol and fentanyl will give rise to a small decrease in ICP (Fitch *et al.* 1969), or unchanged ICP (Miller *et al.* 1975). Consider-ing these experimental studies and the clinical studies of CBF and ICP, neurolept anaesthesia with barbiturate induction has been advocated in surgery for space-occupying intracranial lesions (Marchall 1973, Arner and Gordon 1976). In patients subjected to craniotomy for supratentorial tumors during anaesthesia with thiopentone induction, nitrous oxide 67%, fentanyl and droperidol 0.2 mg/kg during moderate hypocapnia, CBF and $CMRO_2$ were decreased by 35% in comparison with values obtained in normal awake subjects, and the relative CO_2 reactivity was preserved (Cold *et al.* 1988). Another finding was a significant negative correlation between plasma droperidol and $CMRO_2$. In comparison, during anaesthesia with nitrous oxide, halothane 0.5%, fentanyl, hyperventilated to the same level of $PaCO_2$, the preoperative values of $AVDO_2$ were significantly higher, while MABP was at the same level. These findings suggest that halothane anaesthesia in comparison with neurolept anaesthesia give rise to a relative luxury perfusion state (see Fig. 1) (Engberg *et al.* 1989a).

4. Central Analgetics

Morphine

In dogs subjected to incremental doses of morphine up to 1.2 mg/kg a progressive decrease of $CMRO_2$ and CBF to 85% and 45% of control has been found. A single dose of morphine 2 mg/kg had the same effect. These effects were reversed by pain stimulation (Kuramoto et al. 1979) and nalorphine, which initially produce an overshoot in both CBF and $CMRO_2$ (Takeshita et al. 1972). Compatible with these results, CSF pressure is unchanged when morphine is administered to artificially ventilated dogs, but under these circumstances nalorphine increases CSF pressure (Weitzner et al. 1963).

Human studies in healthy volunteers during nitrous oxide and normocapnia have shown that morphine in doses ranging from 1–3 mg/kg, did not change CBF, $CMRO_2$ or CMR-glucose, but decreased MABP and increased CVR, leaving CBF unchanged (Jobes et al. 1977), and the cerebral autoregulation is not affected by morphine (Jobes et al. 1975). Thus, during spontaneous respiration morphine like all central analgetics induces respiratory depression with hypercapnia, which secondarily will decrease CVR and increase CBV, CBF and ICP, leaving $CMRO_2$ unchanged. In dosages given for premedication morphine has little effect on the EEG activity. However, in doses of 1–2 mg/kg, some decrease in frequency might occur and a more frontal distribution of activity might be observed. During nitrous oxide, intravenous morphine will change EEG with desynchronizing, an increase in beta activity and drop out of frontal theta activity (Stockard and Bichford 1975).

Pethidine, gives rise to a small decrease in $CMRO_2$ (Messick and Theye 1969), and like morphine, pethidine given during nitrous oxide anaesthesia provokes excitatory effects on the EEG (Pearcy et al. 1957, Stockard et al. 1972). *Phenoperidine* a derivate of pethidine has been used combined with droperidol for neuroanaesthesia. Given together with droperidol to patients with normal CSF-pathway, ICP is unchanged (Fitch et al. 1969); and in patients with normal arteriography, CBF is unchanged during the combined use of droperidol and phenoperidine (Wilkinson and Brown 1970). However, given alone to patients with severe head injury and intracranial hypertension, an increase in ICP has recently been published (Grummitt and Goat 1984). In another study of traumatic coma, phenoperidine did not change ICP but a significant decrease in MABP and CPP was observed, suggesting that bolus administration of phenoperidine should be avoided in traumatic coma (Bingham and Hinds 1987).

Fentanyl

In experiments with cats, fentanyl induces an increase in CBF and $CMRO_2$ (Nilsson and Ingvar 1965, Freeman and Ingvar 1967). In comparison, studies of venous outflow in dogs during nitrous oxide, indicate a 18% decrease in $CMRO_2$ and a 47% decrease in CBF after fentanyl in doses of 0.006 mg/kg (Michenfelder and Theye 1971). The discrepancy in results might be explained by the fact that fentanyl in some species induces epileptic EEG changes and metabolic activation (DeCastro et al. 1979, Sebel et al. 1981, Safwat and Daniel 1983, Tommasino et al. 1984). Thus, in rats fentanyl in small doses decreases CBF and suppresses the EEG, while higher doses induce epileptic activity and augment CBF as well as $CMRO_2$ (Safo et al. 1983). During epileptic seizure activity the percentage increase in $CMRO_2$ is higher than that of CBF, suggesting that the increase in flow does not follow the metabolic demand, and thus create a state of impeding ischaemia (Carlsson et al. 1982). In rats subjected to fentanyl-induced epileptoid activity a relative hypermetabolism in the limbic system, coupled with a significant reduction of glucose utilization in the rest of the brain, suggests a role of the limbic system in the genesis of seizure activity during fentanyl administration (Tommasino et al. 1984). Although high-dose fentanyl suppresses global $CMRO_2$, studies in rats indicate a failure of fentanyl to prevent the reduction in high energy phosphate and the accumulation of lactate in brain tissue during hypoxia (Keykhah et al. 1988). In dogs, the CO_2 reactivity is unaffected by fentanyl and the drug does not change the autoregulation curve (McPherson and Traystman 1984). In dogs, acute tolerance for the effects of fentanyl-evoked cardiovascular responses occurs within 3 h and rebound withdrawal effects have been observed within 4–5 h (Askitopoulou et al. 1985). In dogs fentanyl, in comparison with halothane induces a reduction of the resistance against resorption of CSF (Artru 1984d).

Human studies of patients with space-occupying lesions indicate that fentanyl injection might increase ICP (Miller et al. 1975). The increase in ICP is secondary to hypercapnia, caused by pulmonary hypoventilation. In contrast, during induction of anaesthesia with thiopentone, followed by fentanyl and controlled hyperventilation ICP is unchanged (Moss et al. 1978) and during induction of anaesthesia with fentanyl 100 µg/kg and diazepam 0.4 mg/kg for cardiac surgery a 25% decrease in CBF was observed, while $CMRO_2$ was unchanged.

High dose fentanyl and sufentanil has been found to be satisfactory for neurosurgery (Shupak et al. 1983, Shupak et al. 1985). However, both forms of narcotic anaesthetics require postoperative respiratory treatment or postoperative administration of nalaxone; furthermore, skilled nursing care must be available for many hours after surgery (Shupak et al. 1985). It is important to know that during prolonged anaesthesia or sedation with fentanyl, the anaesthetic effect might be markedly reduced, suggesting the development of tolerance (Novack et al. 1978, Colpaert et al. 1980, Shafer et al. 1983).

Studies of neurolept anaesthesia (the combined use of droperidol and fentanyl with nitrous oxide) for craniotomy suggest that even a small single dose of nalaxone (1 μg/kg) effectively antagonizes the postoperative respiratory depression without eliminating the immediate postoperative analgetic effect of neurolept anaesthesia (Arnér and Gordon 1976) and recent studies by Wiklund (1986) suggest that the combined use of nalaxone and physiostigmine might reverse sedation and respiratory depression even more effectively. In this respect, studies by Cartwright et al. (1983), indicate that hyperventilation commonly used during neuroanaesthesia significantly increases whole body clearance of fentanyl.

During high-dose fentanyl anaesthesia (50–70 μg/kg) a massive increase in delta power in the EEG has been observed (Sebel et al. 1981). In accordance with the experimental studies, epileptic seizures have been described after fentanyl injection (Rao et al. 1982, Sebel and Bovill 1983). In humans, somato-sensory evoked potentials remain consistently recordable and interpretable during high dose fentanyl anaesthesia (Schubert et al. 1986).

In rats, Sufentanil decreases CBF and $CMRO_2$ and in studies using the open window technique a bolus injection or continuous infusion of sufentanil decreases pial arteries diameter dose-dependently (Lu et al. 1987). Using the venous outflow technique sufentanil increases CBF in the dog, while $CMRO_2$ is unchanged (Milde and Milde 1987). In dogs subjected to inflation of an epidural balloon, sufentanil produced an increase in CBF and ICP and a fall in ATP and phosphocreatine in the area of the balloon (Milde and Milde 1989). In rats high doses of sufentanil provoke epileptic seizure activity (Keykhah et al. 1985), and studies of regional glucose utilization in rats have shown a selective increase in glucose utilization in subcortical limbic nuclei, particularly the amygdala, suggesting that the EEG pattern of seizure activity might reflect subcortical rather than cortical activation (Young et al. 1984).

Clinical studies of EEG under sufentanil anaesthesia have shown high voltage and slow delta wave activity (Bovill *et al.* 1982). Patients undergoing craniotomy with nitrous oxide, isoflurane anaesthesia and subjected to low dose fentanyl or sufentanil infusion show a significantly better cerebral relaxation than without narcotics (Bristow *et al.* 1987). In patients undergoing coronary artery surgery a preliminary study indicate that sufentanil in a dose of 10 μg/kg induces a fall in CBF and $CMRO_2$ and an increase in CVR (Murkin *et al.* 1988).

In dogs, *Alfentanil* induces a dose-related decrease in CBF, the CO_2 reactivity is preserved and the upper limit of autoregulation is at a higher pressure (McPherson *et al.* 1982). In comparative studies in patients with supratentorial tumours, fentanyl did not change lumbar CSF pressure, while sufentanil and especially alfentanil injection were followed by an increase in lumbar CSF pressure, and a decrease in MABP and CPP (Marx *et al.* 1989). In rats subjected to mechanical ventilation, alfentanil increases muscle rigidity, leading to hypercapnia and a significant increase in ICP, which rapidly normalize after non-depolarizing relaxants (Benthuysen *et al.* 1985).

Comparative *clinical studies* of EEG activity during fentanyl, alfentanil and sufentanil injection, indicate activation of EEG after all three drugs, without any significant difference (Ty Smidt *et al.* 1986); however, studies of EEG during induction of anaesthesia with fentanyl, sufentanil or alfentanil have shown that epileptic EEG activity was never detected and there was no EEG evidence of a postictal state in any patient. The same authors emphasize that rigidity is often explosive in onset and mimics epileptic seizures (Smith *et al.* 1987). In a recent study by Marx *et al.* (1988) lumbar CSF pressure was measured in nitrous oxide anaesthetized patients with supratentorial tumours, and the effect of pentanyl, alfentanil and sufentanil on CSF pressure was compared. There was no change in CSF pressure after pentanyl. In contrast, both sufentanil and alfentanil caused peak increases of 89% and 22% respectively. Continuous intravenous alfentanil sedation has recently been proposed in the intensive care of patients subjected to controlled ventilation (Yate *et al.* 1984, Cohen and Kelly 1987, Sinclair *et al.* 1988).

Nalaxone

This n-allyl derivate of oxymorphine is widely used by anaesthetists to reverse opiate-induced respiratory depression. Given intravenously during halothane anaesthesia in dogs, nalaxone does not change CBF or $CMRO_2$ (Artru *et al.* 1980). However, in rats subjected to nitrous oxide or nitrous oxide supplemented with sufentanil, nalaxone induces a significant increase in CBF as well as $CMRO_2$, suggesting that these effects of nalaxone are mediated by opiate receptors and that the depressant effects of N_2O on CBF and metabolism are mediated by opiate receptors (Keykhah *et al.* 1985).

Experimental studies suggest that nalaxone can improve neurological outcome after spinal injury (Faden *et al.* 1981a, Faden *et al.* 1982a, Flamm *et al.* 1982), and nalaxone has been found to improve spinal cord blood flow in the injured region (Faden *et al.* 1981b, Young *et al.* 1981). However, no human trials have been performed, and the clinical use in spinal cord injury has not been established.

In gerbils subjected to carotid occlusion, intraperitoneal injection of nalaxone reverses the stroke for up to 30 min (Hosobuchi *et al.* 1982), and in nalaxone treated stroked animals, CBF is maintained in all cerebral regions above a critical threshold and a substantially better cortical somato-sensory-evoked response recovery has been observed (Faden *et al.* 1982). Although nalaxone in *clinical studies* has been reported to completely reverse neurological deficits in patients with cerebral ischaemia (Baskin and Hosobuchi 1981), no controlled studies are yet available. In this context is to be mentioned that clinical use of nalaxone is not without risk as it can provoke hypertensive reactions (Tanaka 1974, Azar and Turndorf 1979), and even acute pulmonary edema (Flacke *et al.* 1977), Concerning the use of nalaxone in the postoperative period after craniotomy (see fentanyl).

5. Neuromuscular Blocking Agents

Succinylcholine

In *animal experiments* succinylcholine increases ICP (Cottrell *et al.* 1983, Lanier *et al.* 1986). The increase in ICP is also observed after pre-treatment with thiopentone (Thiagarajah *et al.* 1988). In dogs subjected to succinyl-choline injection, CBF increases within a few minutes and this hyperperfusion is accompanied by EEG activation (Mori *et al.* 1973, Lanier *et al.* 1986) and an increase in $PaCO_2$ (Hagan *et al.* 1983). Thus, the increase in CBF and ICP might partly be caused by the metabolic activation and partly by the increase in $PaCO_2$. However, fasciculation in the muscles of the neck, causing stasis in the jugular veins might also be a factor attributing to the increase in ICP (Cottrell *et al.* 1983). In *human studies* without (Halldin and Wåhlin 1959, Marx *et al.* 1962), and with intracranial space-occupying lesions, an increase in ICP has been observed (March *et al.* 1980, Minton *et al.* 1986c). In patients scheduled for craniotomy, induction of anaesthesia with thiopentone pancuronium followed by light enflurane anaesthesia allows successful control of ICP, while the use of succinylcholine for intubation gives rise to an increase in ICP (McLeskey *et al.* 1974).

During induction of anaesthesia pre-treatment with Vecuronium (Minton *et al.* 1986c), or Metocuronium (Stirt *et al.* 1986) attenuates the increase in ICP. Of some interest are studies by White *et al.* (1982) indicating that pre-treatment with succinylcholine before suction procedure in severely injured head injury patients, attenuates the ICP increase better than that obtained with thiopentone 3 mg/kg, lidocaine 1.5 mg/kg intravenously or intratracheally and fentanyl 1 µg/kg.

In patients with cerebral aneurysm succinylcholine injection might provoke hyperkaliaemi (Iwatsuki *et al.* 1980); however, in a recent study patients undergoing early surgery (less than 4 days) following subarachnoid hemorrhage are not at increased risk of development of hyperkalemia (Manninen *et al.* 1989). This reaction has not been observed in patients with brain tumors (Minton *et al.* 1986a).

In rats *d-tubocuranine* induces an increase in CBF, but only when the blood–brain-barrier has been opened with urea. The increase in CBF can be antagonized with cimitidin (Vesely *et al.* 1986). In this context, studies by Gross *et al.* (1981) indicate that histamine might increase CBF only when the blood–brain-barrier is defect. This observation therefore suggests that the effect of d-tubocuranine on CBF is not caused by a direct effect on cerebral vasculature but is effected by histamine, either through a direct vasodilating effect by histamine (Tindall and Greenfield 1973, Martins *et al.* 1980), or as a result of histamine-induced increases in $PaCO_2$ (Weiss *et al.* 1977). In *patients* scheduled for stereotaxic intervention, d-tubocuranine provokes an increase in ICP and a decrease in the average thalamic baseline impedance, suggesting cerebral vasodilatation (Tarkkanen *et al.* 1974). However, the only human study concerning the effect of d-tubo-curanine on CBF, indicates unchanged CBF after d-tubocuranine injection (Wollman *et al.* 1965).

In dogs *Pancuronium* does not affect ICP, CBF, $CMRO_2$ or EEG (Lanier *et al.* 1985).

In cats *Atracurium* does not influence ICP (Giffin *et al.* 1985). However, in the same animal this drug might activate the EEG, without changing CBF or $CMRO_2$ (Lanier *et al.* 1985). Laudanisine, a metabolite of atra-curium provokes epileptic seizures (Mercier and Mercier 1955), and an increase in the concentration of laudanisine has been observed after the use of atracurium (Fahey *et al.* 1984, Boheimen *et al.* 1984). However, in rabbits subjected to cortical application of the epileptigenic drug cefazolin during craniotomy, no increased incidence of seizure activity could be detected in those animals receiving laudanosine at rates sufficient to produce plasma levels of this drug similar to what might be seen following the clinical use of atracurium (Tateishi *et al.* 1989). Accordingly, it has been argued that the concentrations of laudanosine after neuromuscular blockade with atra-curium are incapable of producing neurological or cardiovascular disturb-ances (Chapple *et al.* 1987). *Human studies* indicate that atracurium does not influence ICP (Minton *et al.* 1985, Rosa *et al.* 1986a, Unni and Young 1986).

Like pancuronium, *Vecuronium* does not provoke an increase in ICP in cats (Giffin *et al.* 1986) or in *human studies* (Minton *et al.* 1986b, Rosa *et al.* 1986b).

6. Other Drugs Used in Neuroanaesthesia

Lidocaine

In monkeys, intravenous infusion of lidocaine in subconvulsive doses is associated with diffuse slowing of the EEG activity, and accordingly lidocaine has anticonvulsive properties in experimental epilepsy. On the other hand, lidocaine might elicit epileptic seizures and an arterial threshold concentration has been defined. Like central analgesics, inhalation anaesthetics and hypnotics, lidocaine decreases the utilization of acetylcholine in the brain (Ngai *et al.* 1979).

In dogs, lidocaine in a non seizure-producing dose of 3 mg/kg reduces $CMRO_2$ by 10% after 2–3 min, and $CMRO_2$ is normalized within 5 min. CBF is unchanged. In doses of 15 mg/kg $CMRO_2$ is decreased by 27% and CBF is decreased as well. The metabolic depression is of 20 min duration. During continuous lidocaine infusion until EEG seizures were induced, $CMRO_2$ increased by 12% of control, and CBF increased disproportionately more than the increase in $CMRO_2$. With the determination of seizures $CMRO_2$ returned to control levels within 90 min (Sakabe *et al.* 1974). In rats neither non-seizure or seizure doses of lidocaine cause any reduction in cerebral energy charge or other evidence of increased anaerobic metabolism in the cerebral cortex (Maekawa *et al.* 1981). However, studies using electrodes in the deep cerebral structures (Wagman *et al.* 1967) and studies of regional glucose utilization in rats during EEG activation have shown increased glucose utilization within the limbic system and a simultaneous reduction in glucose utilization in the rest of the brain (Ingvar and Shapiro 1981). Correlated studies of local CBF and glucose utilization have shown an uncoupling of local brain metabolism from blood flow during lidocaine induced subcortical epileptoid discharges in areas recognized to be prone to irreversible damage following prolonged seizure activity (Tommasino *et al.* 1986).

In dogs subjected to deep barbiturate anaesthesia or hypothermia, leading to isoelectric EEG, the addition of lidocaine in doses of 160 mg/kg caused an additional metabolic inhibition of 15–20%, suggesting that lidocaine under these circumstances might provide some protection of the brain (Astrup *et al.* 1981b). In the same studies lidocaine caused reduction in the potassium efflux rate of about 50%, probably by reducing membrane ion permeability in accordance with its local anaesthetic action (Astrup *et al.* 1981a). These results indicate that lidocaine might have some protective effects against cerebral ischaemia, and in experiments in cats subjected to spinal cord injury a certain protective effect has indeed been observed (Kobrine *et al.* 1984, Chang *et al.* 1986), while a protective effect has not been observed in other studies of cats subjected to spinal injury and monitored with spinal evoked responses (Haghighi *et al.* 1987), in rats subjected to near complete global ischaemia (Warner *et al.* 1988), and in cats subjected to MCAO (Shokunbi *et al.* 1986, Gelb *et al.* 1988). However, in a recent study from the same group of cats subjected to MCAO, lidocaine in bolus dosis of 5 mg/kg, followed by continuous infusion gave better preservation of evoked potentials, smaller infarct size and higher CBF in the peripheral region of the infarct (Shokunbi *et al.* 1987). In other studies in cats subjected to cerebral ischaemia induced by air embolism, lidocaine stabilizes blood pressure (Evans *et al.* 1984), and pre-treatment with lidocaine attenuates the intracranial hypertension caused by the air embolism (Evans *et al.* 1989). Furthermore, lidocaine given after the air embolus was also effective in reducing ICP hypertension (Evans and Kobrine 1987). It should be mentioned that studies in dogs during isoflurane anaesthesia, where lidocaine in doses of 15 mg/kg was given have shown a decrease in the concentration of ATP in cerebral tissue, suggesting a detrimental effect of lidocaine on cerebral metabolism (Milde and Milde 1987). However, low concentration of ATP in cerebral tissue before lidocaine administration make this conclusion questionable.

Pain of central origin appears to be affected favourably by intravenous lidocaine, whereas pain of peripheral origin is unaffected, except at blood concentrations which approach toxic values (Boas *et al.* 1982). Lidocaine in clinical doses suppresses the cough reflex (Poulton *et al.* 1979).

During *endotracheal intubation* (Bedford *et al.* 1980b, Hamill *et al.* 1981), scalp incision (Bedford *et al.* 1980a) and suction procedures in neurointensive patients (Donegan *et al.* 1980, White *et al.* 1982, Yano *et al.* 1986), intravenous lidocaine in doses of 1.5 mg/kg inhibits or attenuates the ICP increase. During *induction of anaesthesia* and *intubation*, lidocaine in doses of 1–2 mg/kg attenuates the cardiovascular response of hypertension and tachycardia (Abou Medi *et al.* 1977). In other studies where lidocaine 1.5 mg/kg was given 1,2,3, and 5 min before intubation, lidocaine attenuates the increase in heart-rate and blood pressure only when given 3 min before intubation (Tam *et al.* 1987). However, recent studies indicate that these effects were of minor degree or absent. Thus, lidocaine did not prevent the rate–pressure–product to reach levels which might be dangerous in patients with ischaemic heart disease (Chraemmer-Jørgensen *et al.* 1986), and in a recent study in human by Laurito *et al.* (1987) aerolized or intravenous lidocaine was not found to be more effective than placebo for the control of haemodynamic response to intubation. Furthermore, in a controlled study of patients with intracranial malformation or brain tumors, a dose of lidocaine (1.5 mg/kg) administered 2–3 min prior to intubation with a sleep dose of thiopentone, did not protect against hypertensive pressure peaks in MABP, whereas fentanyl in doses of 6 μg/kg did so (Bachofen 1988).

The effects of other drugs on the pressure response during intubation have been investigated. The beta blocking agent *lobetalol* has recently been investigated in double blind clinical studies. In doses of 0.25 mg/kg within one minute before intubation, this drug will blunt the increase in heart rate and attenuate the increase in blood pressure (Inada *et al.* 1987, Bernstein *et al.* 1987). In doses greater than 0.75 mg/kg it will produce a significant reduction of MABP as well (Leslie *et al.* 1987).

Mannitol and Furosemide

In patients with intracranial hypertension, early studies of intravenous infusion of mannitol showed that this drug effectively reduces ICP (Wise and Chater 1962). After fast intravenous infusion of 0.5–1 g/kg, ICP will be reduced after 2–5 min and the ICP reducing effect will be of hours duration, dependent on dose and infusion rate (James 1980). The osmotic effect of mannitol is dependent on the osmotic gradient in blood (Shenkin *et al.* 1962). A difference in osmotic gradient exceeding 10 mOs always gives rise to a reduction in ICP (Marshall *et al.* 1978), and the decrease in ICP is correlated to the decrease in the water content in brain tissue (Nath and Galgraith 1986).

Animal experiments. Immediately after mannitol infusion in dogs (first 2–3 min), ICP will increase and this effect is correlated to an increase in CBV. The increase in CBV persists for 15 min after mannitol infusion, while the ICP returns to control level within five minutes and continues to decrease (Ravussin *et al.* 1985). Abou-Madi *et al.* (1983) demonstrated that the initial ICP increase can be attenuated when hypocapnia is simultaneously induced. Recent studies from the same group have shown that the initial ICP increase after mannitol is elicited only in animals with normal ICP, while this effect is eliminated in animals with intracranial hypertension (Abou-Madi *et al.* 1987).

In studies of central haemodynamics in rabbits and dogs, mannitol infusion increases blood volume, CVP, wedge pressure and cardiac output, while the concentration of haemoglobin, plasma-natrium and the peripheral resistance will decrease. The blood pressure will change, dependent on the change in cardiac output and peripheral resistance, and often a fall in MABP is observed (Cote *et al.* 1979, Katz *et al.* 1986, Ravussin *et al.* 1986a).

Following mannitol infusion in cats, blood viscosity will decrease immediately, the greatest decrease occurring at 10 min, and at 75 min a rebound increase in viscosity is to be observed. The pial vessel diameter decreased simultaneously, the largest decrease being at 10 min. The changes in pial vessels were interpreted as an autoregulatory process (Muizelaar et al. 1983). Further studies indicate that if the autoregulation is impaired, mannitol infusion will be followed by a decrease in ICP, while ICP is unchanged in studies where the autoregulation is intact (Muizelaar et al. 1984). One of the possible explanations for these findings is based on the indication that autoregulation is mediated through alterations in the level of adenosine in response to changes in the oxygen availability in cerebral tissue. The decrease in blood viscosity after mannitol administration should improve oxygen transport to the brain. When autoregulation is intact, more oxygen leads to the decreased adenosine levels, resulting in vasoconstriction. The decreased resistance to flow from the decreased blood viscosity is then balanced by increased resistance from vasoconstriction, so CBF remains unchanged. In spite of the unchanged CBF, the increase in resistance will decrease CBV and enhance the osmotic dehydrating effect on ICP. When autoregulation is abolished vasoconstriction does not occur in response to increased oxygen availability and an increase in CBF occurs owing to the decrease in viscosity. With a lack of vasoconstriction, the effect on ICP through dehydration is not enhanced so that the resulting decrease in ICP is much smaller (Muizelaar et al. 1984). This hypothesis is based on the assumption that changes in blood viscosity might give rise to compensatory changes in the degree of constriction in cerebral vessels; however, in studies of vessel diameters with the cranial window technique in cats, it was concluded that mannitol in clinically relevant doses does not exert significant constriction on cerebral vessels and mannitol must therefore exert its effect on ICP through its osmotic effect, rather than by a direct effect on cerebral blood volume (Auer and Haselsberger 1987). Accordingly, studies of mannitol infusion at normal intracranial pressure in baboons (Johnstone and Harper 1973) and dogs (Kassell et al. 1982) have shown an increase in CBF, while CBF is unchanged in animals subjected to intracranial hypertension by an epidural balloon (Johnstone and Harper 1973). In the same study mannitol infusion was followed by an increase in $CMRO_2$.

Mannitol has been shown to have beneficial effects on experimental cerebral ischaemia (Little 1978, Watanabe et al. 1979). In studies of experimental cytotoxic edema, mannitol induces a normalization of EEG (James et al. 1978), and studies in rabbits subjected to MCAO have shown that mannitol administration improves CBF in regions of ischaemia and the gradual decline in intercellular pH is prevented (Meyer et al. 1987). In the same ischaemic model using cats, animals receiving mannitol had an improved postischaemic recovery of blood flow (Tanaka and Tomonaga

1987), and in rats subjected to forebrain ischaemia, mannitol considerably ameliorated the ischaemic injury (Sutherland *et al.* 1988). However, other experimental studies of cerebral ischaemia have failed to demonstrate any enhancement of CBF by mannitol (Seki *et al.* 1981, Pena *et al.* 1982) and mannitol treatment in monkeys subjected to MCAO have failed to improve clinical status or decrease infarct size (Pena *et al.* 1982).

Human studies of central haemodynamics in patients undergoing craniotomy have shown that mannitol infusion is followed by an increase in blood volume, CVP, wedge pressure and cardiac output and a decrease in the concentration of haemoglobine, plasma-natrium and the peripheral resistance (Rudehill *et al.* 1983, Brown *et al.* 1986). The concentration of plasma-potassium will decrease after doses of 1 g/kg, and increase after doses of 2 g/kg mannitol (Manninen *et al.* 1987).

Other studies indicate an increase in CBV a few minutes after mannitol infusion (Ravussin *et al.* 1986a) and following mannitol infusion, blood viscosity will decrease for at least two hours, suggesting an enhancement of cerebral microcirculation (Burke *et al.* 1981). Accordingly, studies of cerebral circulation have shown an increase in CBF occurring after 10–20 min and lasting for upto 24 hours (Bruce *et al.* 1973, Mendelow *et al.* 1985, Jafar *et al.* 1986) and a variable increase in $CMRO_2$ (Bruce *et al.* 1973, Jafar *et al.* 1986). In patients subjected to operation for cerebral tumors or aneurysms, mannitol infusion in patients with normal ICP showed a transient but significant increase in ICP, followed by a steady decrease towards values below control. In contrast, patients with intracranial hypertension showed no increase in ICP, which decreased immediately after mannitol infusion (Ravussin *et al.* 1986b). Studies of the effect of mannitol on ICP have shown a dependency on the cerebral autoregulation. Thus, mannitol will only decrease ICP effectively in patients with intact cerebral autoregulation (Muizelaar *et al.* 1984). Other studies suggest that the effect of mannitol is at least partly dependent upon other hemodynamic mechanisms. Thus, patients with CPP > 70 mm Hg responded relatively poorly to mannitol, while ICP decreased in patients with CPP < 70 mm Hg, suggesting that at CPP > 70 mm Hg, the vasoconstriction is already nearly maximum and mannitol is therefore unable to increase resistance further (Rosner and Coley 1987). In neurointensive patients, the volume pressure relationship will improve after mannitol, often without change in ICP (Miller *et al.* 1975).

Preliminary uncontrolled studies indicate that mannitol decreases PaO_2. It is suggested that this decrease reflects interstitial pulmonary edema due to expansion of the extracellular volume and increased hydrostatic pressure (Edle and Smalley 1979). Up to present, controlled studies have not confirmed this observation.

Mannitol and Blood–Brain-Barrier

Osmotic opening of the brain–blood-barrier by infusion of hyperosmolar solutions like mannitol has repeatedly been observed and it has been argued that opening of tight-junctions is the dominant mode of leakage in hyperosmolar opening. For review, see Rapoport and Robinson 1986. The opening of the barrier is independent of energy-producing metabolism and it is supposed that osmotic barrier opening is the result of passive shrinkage of endothelial cells and the surrounding tissue (Greenwood *et al.* 1988). As a result of the impaired barrier function, mannitol diffuse into the cells. Furthermore, brain cells are able to create osmotic active particles. Both factors reduce the transcellular osmotic gradient (Feig and McCurdy 1977, Jennett and Teasdale 1981). After discontinuing the mannitol infusion, the osmotic gradient is reversed because mannitol is excreted through the urine, decreasing the concentration in the plasma. Consequently, the concentration of mannitol is found to be higher in the intracellular compartment as compared with the concentration in the extracellular space. This *rebound phenomenon* give rise to water influx in brain cells, and an increase in ICP (McQueen and Jeanes 1964). The risks of rebound phenomenon indicate that mannitol primarily should be used in single doses in connection with acute ICP increase.

Unlike the changes in *plasma-osmolality*, studies in rabbits subjected to hollow-fiber-plasmapheresis have shown that a decrease in *oncotic pressure* does not increase ICP or increase cerebral water content, suggesting that the presence of blood–brain-barrier may minimize the role of oncotic pressure in determining water movement between the cerebral vessels and brain tissue (Zornow *et al.* 1987). Recently this study has been extended to include rabbits with cryogenic lesions indicating that reduction in oncotic pressure does not play an important role in brain edema formation and that a decrease in osmolality increases water content only in relatively normal brain regions far from the cryogenic lesion (Kaieda *et al.* 1989b). According to the authors, these studies suggest that excessive infusion of even minimally hypo-osmotic fluids, such as lactated Ringer's solution, may be detrimental.

Furosemide induces an inhibition of CSF production and a concurrent reduction of ICP. These changes are thought to enhance the clearance of vasogenic edema (Reulen *et al.* 1977). In cats subjected to cold injury, furosemide effectively reduces the amount of brain edema. In the same study the combination of furosemide, acetazolamide and glucosteroid was even more effective (Long *et al.* 1976), and in dogs subjected to cold injury, furosemide decreases brain water content in normal dogs, but not in nephrectomized dogs, indicating that the effect of furosemide is mediated by diuresis (Marshall *et al.* 1982). In dogs, mannitol and furosemide, when used together, produce a greater and more sustained fall in ICP than mannitol alone (Pollay *et al.* 1983, Roberts *et al.* 1987).

In *human studies* furosemide does not provoke an initial increase in ICP and does not change serum osmolality or electrolytes to the same degree as does mannitol (Cottrell *et al.* 1977). Furosemide decreases CSF formation rate and increases CSF absorption capacity (Sklar *et al.* 1980). In patients with brain tumors or cerebral aneurysms, subjected to a rapid infusion of mannitol (1.4 g/kg) or the same dose combined with furosemide (0.3 mg/kg), brain shrinkage was greater and more consistent with the mannitol plus furosemide, than with mannitol alone, but rapid electrolyte depletion of sodium was observed with the combination of the two drugs, and must be corrected (Schettini *et al.* 1982). In patients with intracranial hypertension, the combination of Ringer's solution and furosemide 250 mg resulted in a definite improvement in general condition occurring after 24 hours and forced diuresis with high doses of furosemide is suggested as treatment of choice for acute cerebral oedema (Thilmann and Zeumer 1974).

7. Drugs Used for Controlled Hypotension

The aims of controlled hypotension in neurosurgical practice are to decrease blood loss during surgery, to provide a dry surgical field and to diminish the risk of intraoperative aneurysm rupture. The benefits of controlled hypotension must be discussed in connection with the risks of development of cerebral ischaemia. For review, see McDowall (1985). Generally, induced hypotension to MABP levels of 50–60 mm Hg, which indicate the lower level of cerebral autoregulation in normotensive subjects is advocated (Strandgaard and Paulson 1984). In chronic arterial hypertension in baboons (Strandgaard *et al.* 1975, Jones *et al.* 1976), and rats (Barry *et al.* 1982) the autoregulation curve is shifted to a higher level and similar findings have been observed in man (Strandgaard 1976). Other studies in baboons have shown that sympathetic vasoconstriction shifts the lower level of autoregulation towards higher pressures (Fitch *et al.* 1975). Accordingly, the hypotensive level under these circumstances must be at a higher level. The lower level of CBF at which cerebral ischaemia threatens is supposed to be about 20 ml/100 g/min. Below this level the content of brain water increases. The increase in brain water is supposed to be a result of an increase in osmotic active molecules produced by anaerobic metabolism; it is a cytotoxic edema and is usually reversible on restoration of MABP (Symon *et al.* 1979). If the degree of ischaemia is more severe so that CBF decreases to the threshold for failure of cell membrane ion haemostasis at about 10 ml/100 g/min, further cytotoxic edema occurs as a result of sodium passage into the cells. This is a cytotoxic oedema as well and it is reversible unless the cell has been severely damaged (Hossmann 1976, Iannotti and Hoff 1983).

Another point of concern is the period immediately after controlled hypotension. If MABP is allowed to increase rapidly or overshoot pre-hypotensive levels resulting in a rebound hypertensive period, plasma proteins and water are forced through the dilated capillary vessels to produce vasogenic edema and it is suggested that the presence within the circulation of vasodilator agents such as sodium nitroprusside or nitroglycerin might accentuate the vasogenic edema (Forster *et al.* 1978).

In neuroanaesthesiological practice, a wide range of intravenous agents have been advocated to achieve an acutely controlled hypotension. Most popular are drugs with a fast effect on systemic vascular resistance and drugs which are metabolized quickly. Presently, sodium nitroprusside

(SNP), nitroglycerin (NG), adenosine triphosphate (ATP) and adenosine (A), all exerting their effects on capacitance and/or resistance vessels are in use. Furthermore, trimethaphan (TMP), a ganglion blocking agent, diltiazem, a calcium blocker and esmelol, a combined alfa and beta blocking agent have been investigated experimentally and in clinical studies. Generally, these drugs exert their effects on preload and afterload. However, baroreflex activity modifies their actions (Chen *et al.* 1982), and studies in neurally deafferented dogs undertaken with constant cardiac output and constant venous pressure using cardiopulmonary bypass, have shown that SNP and ATP-induced hypotension caused flow redistribution from splancnic to extrasplancnic vascular beds, suggesting that redistribution should be taken into consideration in evaluating the hemodynamic changes during induced hypotension (Hoka *et al.* 1987).

Sodium Nitroprusside (SNP)

SNP is a fast acting and potent vasodilator, which predominantly acts on resistance vessels. The vasodilator effect of SNP and related drugs like nitroglycerin are mediated by activation of guanylate cyclase, resulting in an increase in intracellular levels of cyclic GMP (Schultz *et al.* 1977, Katsuki *et al.* 1977), and accordingly aminophylline potentiates the vasodilator effects of SNP (Pearl *et al.* 1983a).

The *toxicity* of SNP is based on its metabolic degradation to cyanide. Cyanide is transformed into the relatively nontoxic thiocyanate, mediated by a hepatic and renal enzyme system, rhodanase, a sulfuryl transferase. The rhodanase enzyme system is a mitochondrial enzyme system, localized close to cytochrome oxidase (Schubert and Brill 1968). In dogs subjected to SNP acutely, cyanide toxicity occurs at doses exceeding 1–1.5 mg/kg (Michenfelder 1977). In human initial dose rates between 0.5–1.5 μg/kg/min are recommended as starting points for very careful titration. Total injected intraoperative dosage should not exceed 3 mg/kg (Tinker and Michenfelder 1976, Vesey *et al.* 1976, Wildsmith *et al.* 1979). Acute toxicity leading to death, during SNP-induced hypotension has repeatedly been described (Jack 1974, Merrifield and Blundell 1974, Davis *et al.* 1975). Thiosulfate is effective in dogs and humans in maintaining a safe blood cyanide concentration, when high doses of SNP are required and without altering the SNP induced hemodynamic effects (Ivankovich *et al.* 1982, Schulz *et al.* 1982). In human hydroxycobalamin has the same effect (Cottrell *et al.* 1978).

In *experimental studies* discontinuation of SNP infusion is followed by an increase in plasma renin activity (PRA), vasopressin and catecholamines (Zubrow *et al.* 1983). These changes are associated with an overshoot or rebound arterial and pulmonary hypertension. Studies in dogs have shown that the rebound hypertension can be prevented partially or completely by pre-treatment with captopril, an inhibitor of angiotensin converting enzyme (Khambatta *et al.* 1982). Experiments with saralasin a competitive inhibitor of angiotensin II have given the same result (Delaney and Miller 1980).

In dogs with artificially induced atelectasis SNP prevents the hypoxic mediated pulmonary vasoconstriction and mediates an increase in Qs/Qt (Colley and Cheney 1977). However, if lung function is normal, experiments in dogs indicate that the shunt fraction is practically unaffected by SNP (Hodges *et al.* 1977).

SNP is a cerebral vasodilator, whether bolus or continuous infusion is used (Ivankovich *et al.* 1976). Using the open window technique in rats, SNP infusion results in a dose-related hypotension, pial arteriolar dilatation and a prolongation of the cerebral microcirculation transit time, without any change in venous diameter (Lu *et al.* 1982). Experiments in cats (Stullken and Sokoll 1975, March *et al.* 1979b), dogs (Todd *et al.* 1980, Prough *et al.* 1983a), have shown an increase in ICP associated with SNP-induced hypotension. As regard the ICP increase, studies in cats with ICP hypertension suggest that slow SNP administration should be preferred under conditions of hypocapnia and hyperoxia (March *et al.* 1979b).

Experiments in rats, cats, dogs and monkeys indicate that the effect of SNP-induced hypotension on CBF may differ with species and experimental conditions. Generally, CBF is unchanged, even at very low levels of MABP (Stoyka and Schutz 1975, Grubb and Raichle 1982, de Guerra *et al.* 1986), or slightly decreased (Carter and Atkinson 1973, Crockard *et al.* 1976, Ishikawa and McDowall 1980). In a recent study in dogs during which CBF, ICP and $CMRO_2$ were measured, SNP alone dilates cerebral capacitance vessels. On the other hand, SNP in combination with isoflurane also dilates resistance vessels. However the latter effect is not reflected by increased CBF or ICP during hypotension because of isoflurane-induced defective autoregulation (Michenfelder and Milde 1988). Studies in cats of the pattern of the brain-surface oxygen tensions during SNP-induced hypotension did not indicate any shift from control, suggesting well maintained cerebral perfusion (Maekawa *et al.* 1979). Other studies by Grubb and Raichle (1982) have shown an increase in cerebral oxygen utilization in baboons during SNP infusion. This effect is supposed to be a disadvantage but concurrent anaesthesia with halothane or other volatile agents is supposed to block the increase in oxygen requirements. The changes in CBF during SNP-infusion are influenced by the capacity of cerebral autoregulation and a left shift of the cerebral autoregulation curve has repeatedly been demonstrated (Stoyka and Schutz 1975, Fitch *et al.*

1976, Michenfelder and Theye 1977, Maekawa *et al.* 1979, Crozier *et al.* 1986). Studies in chronic hypertensive rats have shown that SNP-induced hypotension within the normal range of autoregulation provokes a decrease in both CBF and $CMRO_2$ (Hoffman *et al.* 1982b).

The dilatation of cerebral resistance vessels during SNP-induced hypotension is associated with blood–brain barrier dysfunction (Ishikawa *et al.* 1983). Under the circumstances of dilatation of cerebral vessels, vasopressors and especially norepinephrine may increase ICP critically. A gradual return of MABP to baseline values at the end of a period of induced hypotension seems advisable and the prevention of hypertensive episodes immediately after operation desirable (Fitch 1977).

In dogs, the CO_2 reactivity during SNP induced hypotension is decreased (Gregory *et al.* 1981) or absent (Artru and Colley 1984) but hypocapnia does not intensify anaerobic metabolism even during MABP of about 40 mm Hg (Artru 1986b).

Human studies of SNP-induced hypotension support the experimental findings. SNP infusion is followed by an increase in plasma renin activity, vasopressin (Khambatta *et al.* 1979) and catecholamines (Fahmy and Gavras 1983, Knight *et al.* 1983). These changes are associated with rebound arterial and pulmonary hypertension (Shibutani *et al.* 1983), and resetting of baroreflex sensitivity (Chen *et al.* 1982). During morphine anaesthesia, the baroreflex sensitivity is greater than during halothane anaesthesia, and it is supposed that this difference might explain the difference in dose requirements for SNP in patients anaesthetized with either morphine or halothane (Chen *et al.* 1982).

Human studies have shown that the rebound hypertension can partially be prevented by pre-treatment with captopril, which simultaneously decreases the dose requirement of SNP (Woodside *et al.* 1982, Fahmy and Gavras 1983). Beta blocking agents have the same effect (Lüben and Hempelmann 1982, Khambatta *et al.* 1984).

During SNP-infusion, an increase in Qs/Qt occurs; however, if the lung function is normal, the pulmonary shunt fraction is unaffected (Stone *et al.* 1976).

In patients subjected to SNP-induced hypotension, an increase in ICP was observed when MABP was decreased to 70% of baseline level. If MABP was decreased further, ICP returned towards baseline values and decreased steadily with further decrease in MABP (Turner *et al.* 1977). In other studies including patients with hydrocephalus secondary to fossa posterior tumors, SNP hypotension resulted in an increase in ICP, associated with a decrease in the level of consciousness and complaint of dizziness and headache (March *et al.* 1979a).

In human studies CBF and $CMRO_2$ are unchanged during SNP-induced hypotension to MABP levels of 50–55 mm Hg, or an average decrease in MABP of 32–42% from baseline level (Griffith et al. 1974, Larsen et al. 1982, Henriksen et al. 1983, Pinaud et al. 1989, Bünemann et al. 1987). It is supposed that CBF in normotensive subjects is unchanged until a MABP level of 40–50 mm Hg; however, this level might be influenced by the age of the patients, the level of $PaCO_2$, and the anaesthetic drugs used. In patients with space-occupying intracerebral lesions, cerebral autoregulation is generally lost and ICP may be increased. In this situation higher hypotensive levels not below 55–60 mm Hg are advocated. In chronic hypertension, where a right shift of the cerebral autoregulation curve is present, a higher level of blood pressure should be chosen.

In patients subjected to aneurysm surgery during SNP induced hypotension to MABP 40 mm Hg, CBF increased following discontinuation of SNP infusion and the reestablishment of MABP (Henriksen et al. 1983, Pinaud et al. 1989), suggesting a redistribution of blood flow to the brain in the presence of impaired cerebral autoregulation, or the influence of local brain acidosis (Pinaud et al. 1989).

Nitroglycerin (NG)

In dogs NG predominantly exerts its hypotensive effect on capacitance vessels; however, in comparison with SNP its effect is more inconsistent, the decrease in blood pressure is less rapid, and rebound hypertension is seldom observed.

Like SNP, NP-induced hypotension in cats is associated with an increase in ICP (Rogers et al. 1979b). In dogs, the effect of NP on ICP is dependent on the choice of bolus or continuous infusion. When administered as a continuous infusion the impact on ICP is minimal. In contrast, bolus injection of NG causes a dramatic increase in ICP (Prough et al. 1983b). The same group has compared the effect of SNP and NP in dogs, with or without ICP hypertension by inflation of an epidural balloon and demonstrated that the presence of ICP hypertension diminishes the systemic vasodilating effect of SNP. In contrast, the decrease in MABP and CPP by NG is unaffected by the presence of increased ICP (Prough et al. 1983a).

In dogs CBF and $CMRO_2$ are unchanged until MABP about 40 mm Hg (Colley and Sivarajan 1984, Artru et al. 1986). However, when ICP was increased artificially, the reduction in CPP was sufficient to lead to a decrease in CBF (Rogers et al. 1979). Hypocapnia during NP induced hypotension does not give rise to disturbances of cerebral metabolism or EEG alteration (Artru *et al.* 1986). During NP-induced hypotension by 50%, spinal CBF did not change (Spargo *et al.* 1987). Studies in dogs indicate that NP in comparison with SNP may normalize pulmonary vascular resistance and pulmonary artery pressure in animals with oleic acid pulmonary hypertension. This effect of NP seems to be related to its ability preferentially to dilate the pulmonary vascular bed (Pearl *et al.* 1983b).

Human studies have shown that NG predominantly acts on capacitance vessels (Gerson *et al.* 1982). During continuous infusion of NG a dose-related decrease in MABP and an increase in ICP have been observed. Simultaneously, a decrease in intracranial compliance has been found (Ghani *et al.* 1983). NG administered as a bolus injection produces a marked increase in ICP associated with a decrease in MABP. The maximum decrease occurred 1–2 min following NG, and ICP returned to normal within 2–3 min (Dohl *et al.* 1981). In patients subjected to craniotomy, NG-induced hypotension to MABP averaging 69 mm Hg was followed by an increase in ICP from 14 to 31 mm Hg and a decrease in CPP from 90 to 38 mm Hg, suggesting dilatation of capacitance vessels within the relatively non-compliant cranial cavity (Cottrell *et al.* 1980).

In patients with normal lung function, NG like SNP-induced hypotension is associated with an increase in pulmonary shunting and a decrease in pulmonary artery pressure (Casthely *et al.* 1982).

Concerning the effect on central hemodynamics, studies in patients subjected to aneurysm surgery indicate that NG appears to be inferior to SNP for the production of induced hypotension because of a greater decrease in cardiac index and mixed central venous saturation produced by NP (Maktabe *et al.* 1985).

Trimethaphan (TMP)

TMP produces a ganglionic blockade, resulting in a decrease in both arteriolar and venous tone. TMP causes neuromuscular blockade (Nakamura *et al.* 1980). Recent findings indicate that TMP acts on the motor endplate but unlike d-tubocurarine TMP does not interact with the recognized site of acetylcholine receptors; the action of TMP appears to be associated with the blockade of end plate ionic channels (Nakamura *et al.* 1988).

Studies in cats and dogs have demonstrated that TMP and especially rapid induction of hypotension produces a significant but transient increase in ICP (Stullken and Sokoll 1975, Bunegin *et al.* 1984, Karlin *et al.* 1988).

During TMP-induced hypotension in dogs, the cerebral autoregulation curve is shifted to the left, and CBF is maintained until MABP 50 mm Hg (Fitch *et al.* 1976). In comparison with SNP and NG, studies in cats indicate that CBF is significantly lower during TMP-induced hypotension (Ishikawa and McDowall 1980, Gregory *et al.* 1981) and studies in the same animal have shown that the CO_2 reactivity during TMP-induced hypotension is abolished (Gregory *et al.* 1981, Artru and Colley 1984). Other studies in dogs have shown that hypocapnia does not intensify the cerebral impact of TMP-induced hypotension on CBF (Artru 1986b). In baboons rendered hypotensive with SNP, TMP or acute hemorrhage with a 50% decrease in MABP, CBF decreased by about 20%. In the same study an increase in cerebral oxygen utilization was observed in animals subjected to SNP or hemorrhage but not during TMT induced hypotension. The authors suggest that the increase in cerebral oxygen uptake observed with SNP and during hemorrhage is due to stimulation of the sympathetic nervous system with release of catecholamines (Grubb and Raichle 1982).

Several studies concerning the cerebral effects of TMP, SNP and NG indicate that TMP is inferior to the other two drugs mentioned. Thus, the pattern of brain surface oxygen tension is moved to lower values during TMT-induced hypotension; an observation not found during SNP-induced hypotension (Maekawa *et al.* 1979). It has been demonstrated that electrical activity as measured by the cerebral function monitor is depressed at higher values of MABP during TMP compared with NP induced hypotension (Ishikawa and McDowall 1980). Furthermore, it has been demonstrated that the MABP threshold for K^+ escape into the extracellular fluid in the central cortex is higher during TMP-induced hypotension than during hypotension induced by NP; and extracellular fluid acidosis is less severe during NP compared with TMP induced hypotension (Morris *et al.* 1983). Other studies in dogs subjected to TMP or SNP induced hypotension to 50 mm Hg, have shown metabolic alterations by TMP (Michenfelder and Theye 1977). On the other hand, studies in baboons with intracranial hypertension provoked by inflation of an epidural balloon indicate that moderate hypotension by SNP (20% fall in blood pressure) in comparison with NG and TMP provokes higher ICP levels during inflation and exflation of the balloon and that the CO_2 reactivity is more reduced (Hartmann *et al.* 1989).

Human studies indicate that TMP in comparison with SNP-induced hypotension results in less of an increase in circulatory catecholamines, and no activation of plasma-renin (Knight *et al.* 1983).

Only a few clinical studies concerning the effect of TMP on ICP have been published. Studies in patients undergoing neurosurgery during the induction of deliberate hypotension using either SNP or TMP indicate that

TMP does not usually produce ICP changes except when intracranial compression is severe (Turner et al. 1977). However, the use of TMP has waned recently, partly owing to comparative experimental studies of TMP, SNP and NG, indicating that TMP is inferior to the other drugs and partly owing to clinical studies indicating that other drugs are to be preferred. Thus, it has been demonstrated that the cerebral electrical activity as monitored by cerebral function monitor is depressed at higher values of MABP during TMP compared with SNP-induced hypotension (Thomas et al. 1985).

The use of SNP 25 mg and TMP 250 mg mixtures for deliberate hypotension has recently been proposed. In a recent report, blood cyanide levels were significantly lower with this mixture than with SNP alone (Wildsmith et al. 1983), and studies in human indicate that the use of this mixture allows for a more rapid decrease of MABP, lower dose requirements for SNP and TMP and diminished or absent reflex sympathetic activation (Fahmy 1985, Fahmy et al. 1986).

Adenosine Triphosphate (ATP) and Adenosine (A)

ATP given intravenously in dogs is rapidly degraded to A during transpulmonary passage and the hypotensive effect of ATP is related to the arterial adenosine concentration (Sollevi et al. 1984b). ATP and A-induced hypotension is characterized by rapid induction (Fukunaga et al. 1982, Lagerkranser et al. 1984, Lagerkranser et al. 1985), and a stable blood pressure level without tachycardia or posthypotensive rebound hypertension (Satinover et al. 1981, Fukunaga et al. 1982, Kassell et al. 1983a, Sollevi et al. 1984b). Accordingly, plasma renin activity is not increased during or following ATP or A-induced hypotension (Tagawa and Vander 1970, Lagerkranser et al. 1984).

Dipyridamole pre-treatment decreases dose requirements by its inhibitory effect on A uptake (Kolassa and Pfleger 1975, Wu and Phillips 1982), and is recommended during A-induced hypotension (Sollevi et al. 1984a, Lagerkranser et al. 1988).

In rats, pigs, cats and dogs, adenosine is a potent dilator of cerebral vessels (Berne et al. 1974, Wahl and Kuschinsky 1976, Morii et al. 1986, Stånge et al. 1989). This effect is blocked by theophylline (Wahl and Kuschinsky 1976). In rats, the concentration of A in the brain increases during hypoxia (Winn et al. 1981), hypotension (Winn et al. 1980b), ischaemia (Berne et al. 1974, Winn et al. 1979) and during bicuculline-induced seizures (Winn et al. 1980a). Recent studies indicate the presence of

A receptors (Huang and Drummond 1979) and other studies suggest that A is involved in the maintenance of resting cerebrovascular tone and that A has a paramount role in the regulation of CBF (Berne *et al.* 1974, Forrester *et al.* 1979, Winn *et al.* 1981, Phillis *et al.* 1987). Osmotic disruption of the blood–brain barrier does not affect the vasodilatory effect of ATP (Forrester *et al.* 1979).

In baboons, ATP induced hypotension produces an initial CBF increase followed by a progressive decrease as MABP decreases to 40 mm Hg or 60% of the baseline value (Van Aken *et al.* 1984a). Other studies in rats (Satinover *et al.* 1981) and dogs (Kassell *et al.* 1983a, Kien *et al.* 1987), indicate unaltered CBF unless MABP is markedly decreased. In dogs subjected to A-induced hypotension of 61% decrease in MABP rCBF were heterogenous, leaving the cerebral cortex and corpus callosum most affected and the brain stem least (Kassell *et al.* 1983b). In the same animal, ATP-induced hypotension to MABP 40 mm Hg induces a CBF decrease by 54–65% and under these circumstances, the cerebral circulation was inadequate to meet unchanged cerebral oxygen demands, resulting in anaerobic metabolism with accumulation of lactate (Newberg *et al.* 1985). In rats, measurements of brain oxygen tensions during SNP, deep isoflurane or 2-chloroadenosine indicated impairment of oxygenation during 2-chloroadenosine induced hypotension, but not during deep isoflurane or SNP-induced hypotension (Longnecker and Seyde 1984). Using the microsphere technique, the same authors have studied rCBF but did not find any differences between the three drugs (Monk *et al.* 1987), unless hemorrhage was instituted during the hypotensive period. In this case, rCBF was best preserved during SNP hypotension (Sperry *et al.* 1987). However in sheep subjected to A or SNP induced hypotension to MABP 30 mm Hg brain surface oxygen tensions were decreased equally (Laycock *et al.* 1986), and in rabbits subjected to ATP or SNP-induced hypotension to MABP 50 mm Hg for 3 hours, ATP in comparison with SNP caused minimal metabolic aberration and all animals survived (Chiache *et al.* 1983). Studies in pigs and baboons subjected to A-induced hypotension indicate an increase of CBF in the early post-hypotension period unrelated to rebound hypertension (van Aken *et al.* 1984b, Eintrei and Carlsson 1986, Stånge *et al.* 1989). It is supposed that A may have an effect on the cerebrovascular tone even after termination of its administration but the mechanism behind this effect is unknown (Stånge *et al.* 1989).

In dogs, pigs and monkeys $CMRO_2$ is unchanged during ATP or A-induced hypotension (Kassell *et al.* 1983b, Van Aken *et al.* 1984a, Newberg *et al.* 1985, Stånge *et al.* 1989). However, in baboons, intracarotid ATP increases $CMRO_2$, and this effect was attributed to an elevation of the cellular cyclic AMP level (Forrester *et al.* 1979). In rats, the combination of hypocapnia to $PaCO_2$ 20 mm Hg, and A-induced hypotension to MABP 50 mm Hg leads to a homogenous decline of glucose utilization, the decrease ranging from 20 to 40%. It is supposed that the decrease of glucose utilization may be the result of synaptic inhibition, probably because of impairment of the presynaptic coupling between excitation and transmitter release (Waaben *et al.* 1989).

The effects of A and ATP on ICP have been studied repeatedly. In the study by Thiagarajah *et al.* (1983), cats were subjected to SNP or ATP-induced hypotension to MABP 38% of control. In animals with normal ICP, ATP did not increase ICP, while ICP increased by 86% during SNP. Similar results were obtained in cats with ICP hypertension due to an inflation of an epidural balloon. In dogs, with normal ICP, ATP induces an increase in ICP; however, if ICP is increased by inflation of an epidural balloon prior to ATP administration, the initial ICP increase is followed by a decrease in ICP (Puchstein *et al.* 1983). Similar results have been obtained in dogs by van Aken *et al.* (1984b), who also found an ATP-induced decrease in intracranial compliance as evaluated by the volume-pressure curves.

Studies in rats subjected to hypoxia have shown that ATP has a protective effect (Kraynack *et al.* 1981).

In *human studies* ATP and A-induced hypotension is characterized by rapid induction, stabilized hypotensive level and lack of rebound hypertension (Fukinaga *et al.* 1983, Lagerkranser *et al.* 1988); accordingly, A has recently been introduced as an agent to induce controlled hypotension during aneurysm surgery (Sollevi *et al.* 1984a, Öwall *et al.* 1984). Studies of central hemodynamics in patients undergoing cerebral aneurysm surgery or an operation for peripheral vascular disease indicate that A-induced hypotension produces a hyperkinetic circulation in the systemic as well as in the myocardial vascular bed. (Öwall *et al.* 1988a, Öwall *et al.* 1988b).

In patients undergoing surgery for cerebral aneurysm in neurolept anaesthesia and subjected to A-induced hypotension with MABP decrease to 46 mm Hg, systemic vascular resistance was reduced by 61%, but CBF did not change. $AVDO_2$ decreased by 28% and $CMRO_2$ by 17%. In the posthypotensive period MABP was increased by 10% and CBF by 15%. It was concluded that A-induced hypotension at these levels did not unfavourably affect cerebral oxygenation and even might offer a protective effect by reducing oxygen demand. The slight CBF increase in the posthypotensive period was supposed to be secondary to an increase in MABP together with a blunted cerebral autoregulation (Lagerkranser *et al.* 1988). From the same group, studies by Positron emission tomography in patients with *a-v* malformation have shown that A produces marked cerebral vasodilation in normocapnic subjects and that this response can be counteracted by hypocapnia (Sollevi et al. 1987).

Recent studies in cats subjected to MCAO have suggested a cerebral protective effect of the beta-adrenergic receptor blocking agent propranolol (Little et al. 1982, Standefer and Little 1986). Esmolol, an ultrashort acting cardioselective beta blocker has been studied experimentally and clinically. In dogs subjected to esmolol infusion during hemorrhage, hypovolemia and under circumstances of low intracranial compliance, a decline in CBF appears to be related to a decrease in CPP, suggesting an impairment of cerebral autoregulation (Bunegin et al. 1987). In patients subjected to surgery for arterio-venous malformation in isoflurane anaesthesia, esmolol in doses of 100–300 mcg/kg/min reduces MABP by 15% from baseline level, without associated reflex tachycardia or rebound hypertension (Ornstein et al. 1986). During SNP-induced hypotension esmolol reduces plasma renin activity dose-dependently and prevents rebound hypertension (Edmondson et al. 1988) and in comparison with SNP, other studies indicate that esmolol-induced hypotension is not associated with reflex tachycardia or rebound hypertension (Ornstein et al. 1987).

In experimental studies labetalol (a combined alfa and beta adreno-ceptor blocker) when added to isoflurane for induction of hypotension results in a decreased requirement for isoflurane. In comparison with deep isoflurane-induced hypotension CBF is unchanged, but a restoration of renal and hepatic blood flow to values observed during normotensive anaesthesia is observed (Durieux et al. 1988). In patients with cerebral disorders subjected to craniotomy with SNP-induced hypotension labetalol permitted withdrawal or substantial reduction of SNP dose and significantly improved CPP and reduced ICP compared with SNP therapy (Orlowski et al. 1988). Until now, no clinical studies of cerebral circulation and metabolism during labetalol-induced hypotension are available.

Diltiazem, a calcium blocker possessing vasodilatating properties has been studied experimentally. In cats with normal ICP and subjected to 30% decrease in MABP, diltiazem did not increase ICP; however, in animals with artificially raised ICP by inflation of an epidural balloon, diltiazem-induced hypotension was associated with an ICP increase (Thiagarajah et al. 1986). In dogs with or without ICP hypertension by mock CSF infusion, diltiazem-induced hypotension (40% decrease in MABP) increased ICP very little; however, serious cardiac disturbances occurred (Mazzoni et al. 1985).

Under certain circumstances the calcium blocking agents nifedipine and nicardipine have cerebral protective effects; consequently both drugs have been proposed for controlled hypotension. In experimental studies and in human with arterial hypertension nifedipine produces an increase in ICP

(Giffin *et al.* 1983, Tateishi *et al.* 1988). In patients subjected to maxillofacial surgery with halothane or isoflurane anaesthesia, nifedipine is an effective drug to use for controlled hypotension provided reflex tachycardia is prevented (Spiss *et al.* 1985). In humans without intracranial disorders *nicardipine* induces a rise in ICP (Nishikawa *et al.* 1986). Human studies of CBF during nifedipine- or nicardipine-induced hypotension are not available. However, experiments in dogs subjected to nicardipine-induced hypotension to MABP 40 mm Hg indicate a significant reduction in rCBF which persists despite return of normal blood pressure (Lam *et al.* 1988).

Experimental and clinical data concerning the use of *isoflurane* and barbiturate anaesthesia in deliberate hypotension have been described in Chapters 2 and 3.

Conclusion

Comparative studies of central hemodynamic indicate that adenosine is superior to SNP, NG and TMP as it improves cardiac output and preserves CVP. Furthermore, A is a fast acting, non-toxic drug and tachyphylaxia or rebound hypertension are not observed. However, A like SNP and NG is a cerebral vasodilator and may give rise to ICP hypertension and a dangerous decrease in CPP. A certain degree of brain protection has experimentally been found in A treated animals and in clinical studies a small decrease in $CMRO_2$ has been observed. In comparison with isoflurane, A-induced hypotension does not imply any significant advantages and in European countries adenosine is not commercially available. Isoflurane-induced hypotension is associated with a profound suppression of $CMRO_2$ and probably better brain protection and therefore might be considered more safe.

Experimental studies of calcium blocking drugs and combined alpha and beta blocking agents are promising, but only a few human studies are available and the effects of these drugs on cerebral circulation and metabolism both experimentally and clinically are rare.

8. Human Studies of CBF and Metabolism During Craniotomy

The Intravenous Modification of the Kety and Schmidt Method

During craniotomy externally placed detectors over the skull are almost incompatible with the surgical procedure. The use of the Kety Schmidt procedure for measurement of global CBF and $CMRO_2$ is therefore a possibility. Although this method is cumbersome it has the advantage that samples of blood for CBF and $AVDO_2$ are withdrawn from the same vascular bed.

The classical method described by Kety and Schmidt (1945) is based on the mean transit time for tracer molecules traversing the brain. Generally CBF is measured during a saturation period of 10 to 15 min of inhalation of an inert gas. The methodology of the inert gas method has been described with nitrous oxide (Kety and Schmidt 1948), 85-Kr (Lassen and Munck 1955) and 133-Xe (Høedt-Rasmussen 1967).

An intravenous modification based on the desaturation period instead of the saturation period has been developed for studies of CBF and $CMRO_2$ during craniotomy. In awake patients subjected to craniotomy for supratentorial tumors the mean values $\pm SD$ averaged 47 ± 6 ml and 3.3 ± 0.7 ml O_2 (Astrup et al. 1984). These values are in close correspondence with expected values for global CBF and metabolism as measured by Kety and Schmidt (1948).

Material

By use of an intravenous modification of the method described by Astrup et al. (1984), a series of studies of CBF have been performed in patients subjected to elective craniotomy for supratentorial brain tumors. Before anaesthesia, all patients were alert and undergoing treatment with steroids. During craniotomy, two studies of CBF have been performed, one before surgical incision about one hour after induction of anaesthesia and the other one hour later after opening of dura during surgical evacuation of the tumor, or dissection. During these phases, relatively stable levels of MABP and $PaCO_2$ were obtained and during the second flow measurement after opening of the dura, MABP was identical with the CPP.

Method

After induction of anaesthesia, the internal jugular vein contralateral to the side of the surgery was punctured from the anterior side of the neck about 1 cm lateral to the internal carotid artery, and a 16 G catheter (Certofix, Braun) was introduced by the Seldinger technique. The tip of the catheter was placed at the base of the skull and the position of the catheter was controlled by X-ray. 133-Xe (about 100 MBq dissolved in 30 ml saline) was injected intravenously during a period of 20 min. During the first 3–5 min the infusion rate was 80–100 ml/h, during the next 15–17 min the infusion rate was adjusted to 20 ml/h. Two ml samples of arterial and jugular venous blood were withdrawn simultaneously from the catheters at −1.0, 1, 2, 3, 4, 5, 7, 9, 11, 13, 15, 20, 25 and 30 min during the desaturation period. Sample radioactivity was counted in a well scintillator and the arterial and venous desaturation curves were drawn as shown in Fig. 8. The brain 133-Xe concentration at the start of the desaturation was estimated from the jugular venous blood concentration at time zero. The amount of 133-Xe released by the brain during the 30 min desaturation period was

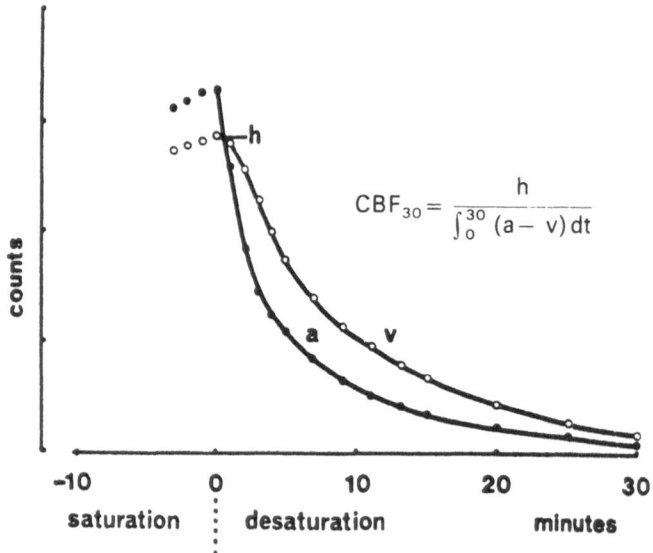

$$CBF_{30} = \frac{h}{\int_0^{30} (a - v)\,dt}$$

Fig. 8. The Kety and Schmidt technique of measuring global cerebral blood flow (CBF) from the 133-Xenon desaturation curves obtained by sampling arterial (*a*) and internal jugular venous (*v*) blood and using the "height over area" formula. The saturation period was obtained with injection of 133-Xenon (3 mCi dissolved in 20 ml saline injected over 20 min). The samplings of blood were obtained in 2 ml air-tight syringes. Sample activity was counted in a well counter

estimated as the area between the arterial and venous desaturation curves by planimetry. The average CBF through the brain during the desaturation period (CBF-30) was calculated as the brain 133-Xe concentration at the start of the desaturation and the 133-Xe arterio-venous difference using the height over area formula of Kety and Schmidt: CBF = height × lambda/area ml/100 g/min, where lambda represents the brain to blood partition coefficient for Xenon. In accordance with Høedt-Rasmussen (1967) the partition coefficient (lambda) was corrected for changes in hemoglobin. Lambda values for the whole brain were used. The $CMRO_2$ was calculated as the product of global CBF and the average of two arterio-venous oxygen differences measured at time zero and 5 min after desaturation. Oxygen tension was measured with ABL1, ABL3 (Radiometer, Denmark) and oxygen saturation with OSM_2 (Radiometer, Denmark). Hemoglobin was measured separately.

The use of this method presumes that equilibrium of Xenon is reached during the saturation period of 20 min, and that the perfusion in the different regions of the brain is homogeneous. In the study by Mc Henry (1967) with 85-Kr as inert tracer a 7 min saturation period was used but as stated later on by Lassen and Klee (1965), this period is too short and consequently in the present study a 20 min saturation period has been used. Another factor of concern is the duration of the desaturation curve; especially in regions where an extremely low CBF is present, a 30 min desaturation curve might be insufficient. Sapirstein and Ogden (1965) have criticized the inert gas method and have shown that quantitatively and qualitatively erroneous results may be obtained by the use of this method for the estimation of CBF. The result should be an overestimation of blood flow because the arterial and venous saturation and desaturation curves never reach each other during the period of investigation.

In a study of CBF and $CMRO_2$ during continuous althesin infusion (Bendtsen et al. 1985), where the intravenous modification of Kety and Schmidt was applied, very low flow values were obtained. 3 mCi Xenon-133 dissolved in about 20 ml saline was injected with a rate of 80 ml/h for 3 minutes and 20 ml/h in 17 minutes in a peripheral vein. After 20 min the activity in 2 ml samples of arterial blood averaged 10000 counts/min, while the count rate in jugular venous blood averaged 8000 counts/min and the activity in jugular venous blood after 30 min averaged 910 counts/min. The calculated CBF values using 10, 15, 20, 25 and 30 min desaturation periods are shown in Fig. 9 for 10 paired studies in 10 patients. As indicated, after a 15 min desaturation period, the CBF values were stabilized in studies with low CBF, indicating that the desaturation curves at this time were practically monoexponential. Consequently, a 30 min desaturation period was supposed to be sufficient to obtain reliable CBF results.

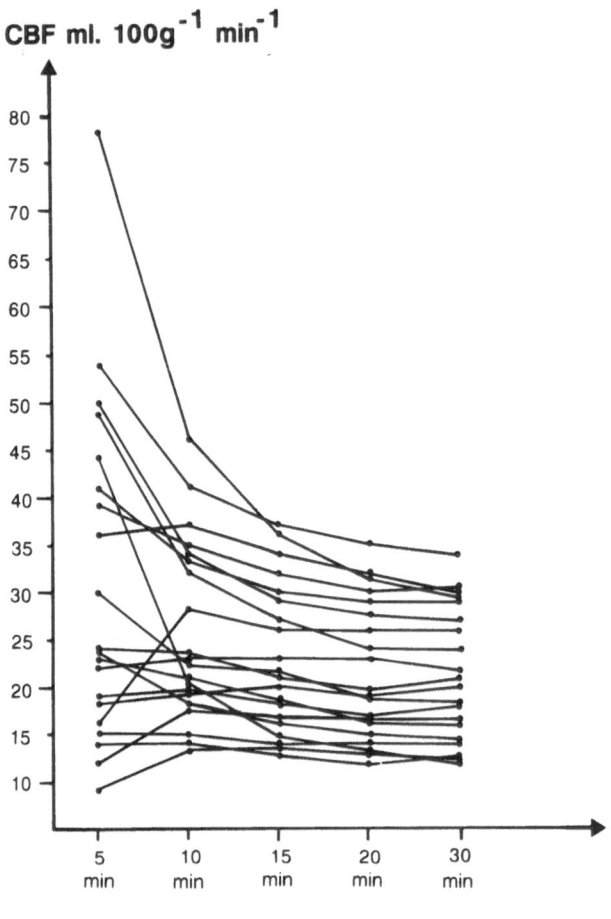

Fig. 9. Correlation between desaturation period in min and CBF calculated by the "height over area" formula using the Kety and Schmidt technique. During the saturation period 3 mCi 133-Xenon dissolved in 20 ml saline was injected i.v. 20 studies of CBF are included. The CBF measurements were obtained in 10 patients subjected to paired studies of CBF during craniotomy for supratentorial cerebral tumors. The patients were anaesthetized with continuous althesin infusion, nitrous oxide 67% and fentanyl. At CBF (30 min) below 25 ml/100 g/min a desaturation period of 15–20 min is sufficient to obtain reliable CBF values

Other methodological errors shall be mentioned, including contamination by central venous blood (Murray *et al.* 1978, Steinbach *et al.* 1976), and blood from extracerebral regions (Lassen and Lane 1961). In the present study, these errors were thought to be eliminated by visualizing the placement of the catheter by X-ray, and by carefully and slow withdrawal of venous blood into airtight syringes. Furthermore, patients in the prone and lateral position were excluded from the study.

Table 1. Indicate the changes in cerebrovascular resistance (CVR), cerebral blood flow (CBF), cerebral metabolic rate of oxygen (CMRO$_2$), electroencephalogram (EEG), intracranial pressure (ICP), mean arterial blood pressure (MABP) and cerebral perfusion pressure (CPP) after 2 MAC halothane, enflurane or isoflurane, 67% N$_2$O, a bolus injection for induction of sleep of barbiturate, althesin, etomidate, propofol, diazepam, midazolam and ketalar; the effects of central analgetics equivalent to 10 mg morphine are indicated; and a bolus dosis of the muscle relaxants for tracheal intubation

	CVR	CBF	CMRO$_2$	EEG	ICP	MABP	CPP
Halothane (h)	− −	+ + +	−	slowing	+ + +	− −	− − −
Enflurane (h)	− −	+ +	− −	slowing seizures	+ +	− −	− −
Isoflurane (h)	− −	+	− − −	slowing burst-suppr.	+	− − −	− − −
N$_2$O (h)	−	+	+	slowing	+ +	0	− −
Barbiturate (h)	+ + +	− − −	− − −	slowing isoelectric	− − −	− −	+
Althesin (h)	+ + +	− − −	− − −	slowing isoelectric	− − −	−	+
Etomidate (h)	+ + +	− − −	− − −	slowing isoelectric	− − −	−	+
Diazepam (a)	?	−	0	slowing	0	−	−
Midazolam (h)	+ +	− −	− −	slowing	− −	−	−
Propofol (h)	+	− −	− −	slowing	− −	− − −	−
Ketamine (h)	− − −	+ + +	+ +	activation	+ +	+ + +	+ +
Morphine (h)	0	−	−	slowing	0	−	−
Fentanyl (a)	+ + +	− − −	− −	slowing activation	−	−	−
Alfentanil (a)	?	− −	+ ?	slowing activation	0	−	?
Sufentanil (h)	+ +	− −	− − + ?	slowing activation	?	−	?
Droperidol (a)	+ + +	− − −	− −	slowing	− −	− − −	− −
Succinylch. (a)	?	+ +	?	activation	+ +	0	−
Pancuronium (a)	0	0	0	unchanged	0	0	0
Vecuronium (h)	?	?	?	unchanged	0	0	0
Alcuronium (h)	?	?	?	unchanged	0	·0	0

−, − − and − − − indicate slight, moderate and pronounced decrease. +, + + and + + + indicate slight, moderate and pronounced increase. 0 indicate no effect. (h) and (a) indicate human and animal studies respectively. Where only (a) is indicated, no human studies are available.

Anaesthetic Procedure

CBF and $CMRO_2$ were studied during the following anaesthetic procedures: *Study 1:* Continuous althesin infusion, nitrous oxide, fentanyl (Bendtsen *et al.* 1985); *Study 2:* Continuous etomidate, nitrous oxide, fentanyl (Cold *et al.* 1985); *Study 3:* Pentobarbitone induction, halothane 0.5%, nitrous oxide, with and without thiopentone loads (8 mg/kg) (Astrup *et al.* 1984). *Study 4:* Neurolept anaesthesia with droperidol 0.2 mg/kg, fentanyl, nitrous oxide (Cold *et al.* 1988); *Study 5:* Continuous midazolam, nitrous oxide, fentanyl (Knudsen *et al.* 1989); *Study 6:* Halothane 0.45 and 0.9% supplemented with nitrous oxide 67 and fentanyl (Madsen *et al.* 1987a); *Study 7:* Enflurane 1.0 and 2.0%, supplemented with nitrous oxide 67% and fentanyl (Madsen *et al.* 1986). *Study 8:* Isoflurane 0.75 and 1.5% supplemented with nitrous oxide 67% and fentanyl (Madsen *et al.* 1987b).

In all studies nitrous oxide 67% was used. All patients were subjected to moderate hyperventilation the aim being $PaCO_2$ between 3.5 and 4.5 kPa (26–34 mm Hg) and PaO_2 above 13 kPa (98 mm Hg). A servo 900B ventilator with end-expiratory CO_2 analyser was used. During the anaesthesia and the CBF measurements mean arterial blood pressure (MABP) was continuously monitored by intra-arterial approach (left radial artery). Rectal temperature was measured repeatedly and during each CBF measurement. More detailed information concerning the patients, materials, anaesthetic procedures and results are presented as follows:

Study 1 (Continuous Althesin)

Patients: 10 patients (mean age 59 years, range 37–75 years) with supratentorial cerebral tumors with midline shift less than 10 mm, as judged by CT scanning or arteriography were included. Before the induction of anaesthesia all patients were awake and were receiving treatment with steroids.

Anaesthesia: One hour before induction of anaesthesia the patients were premedicated with pentobarbitone 2 mg/kg i.m., mepyramine 50 mg i.m. and cimitidine 400 mg by mouth. Anaesthesia was induced with an infusion of althesin at a rate of 1 ml/kg/h and fentanyl 0.2 mg i.v. Pancuronium 0.10–0.15 mg/kg was given to produce neuromuscular blockade, manual hyperventilation was applied and tracheal intubation performed. From the point of intubation until the end of the first CBF measurement, the anaesthesia was maintained with a continuous infusion of althesin 0.2 ml/kg/h, 67% nitrous oxide in oxygen and fentanyl in doses of 0.1 mg. In five patients the rate of althesin was 0.2 ml/kg/h throughout the study and CBF was measured on two occasions. In four patients the althesin infusion rate was increased to 0.5 ml/kg/h 30 min before the second CBF determination. In one patient, the infusion of althesin was unintentionally disrupted 30 min before the second CBF measurement.

Results: During continuous althesin infusion (althesin 0.2 ml/kg/h), the first CBF averaged 22 ml/100 g/min, and during the second study 20 ml/100 g/min. $CMRO_2$ averaged 1.8 and 1.9 ml O_2/100 g/min respectively. No significant changes in $PaCO_2$ and MABP were disclosed (Table 2).

In four patients the althesin infusion rate was increased from 0.2 to 0.5 ml althesin/kg/h. During the first CBF measurement CBF and $CMRO_2$ averaged 27 ml/100 g/min and 1.9 ml O_2/100 g/min and during the second CBF measurement 20 ml/100 g/min and 1.5 ml/O_2/100 g/min. The results are presented in Table 3.

Study 2 (Continuous Etomidate)

Patients: Fourteen patients (mean age 62 years, range 41–73 years) with supratentorial cerebral tumors, with shift of midline structures up to 10 mm at CT scanning were included. Before induction of anaesthesia all patients were awake, and were receiving treatment with steroids.

Anaesthesia: For premedication, pentobarbitone 2 mg/kg i.m. was used one hour before induction. For induction of anaesthesia, an average dose of 0.28 mg/kg etomidate was given i.v., and followed by fentanyl 0.2 mg and pancuronium 0.10 mg/kg. After manual hyperventilation, intubation was performed. Maintenance of the anaesthesia was performed with nitrous oxide in oxygen (67%). Fentanyl was supplemented in doses of 0.1 mg/70 kg body weight every 30–60 min. Neuromuscular blockade was monitored with train-of-four stimulation, and to provide sufficient relaxation, pancuronium was given in a dose of 1–2 mg. In seven patients, the dose of etomidate was adjusted to 0.5 μg/kg/h throughout the anaesthesia. In another seven patients, the etomidate infusion rate was increased to 1.0 μg/kg/h after termination of the first CBF run.

Results: During unchanged and continuous etomidate infusion rate of 0.5 μg/kg/h, CBF and $CMRO_2$ averaged 25 ml/100 g/min and 2.3 ml O_2/100 g/min during the first CBF run, and 22 ml/100 g/min and 2.2 ml O_2/100 g/min during the second CBF measurement (Table 2).

In the group of 7 patients where the etomidate infusion rate was increased from 0.5 to 1.0 μg/kg/min between the first and second CBF run, the average values of CBF were 35 and 23 ml/100 g/min ($P < 0.05$), and the average values of $CMRO_2$ were 2.5 and 1.8 ml O_2/100 g/min ($P < 0.05$) (Table 3).

Table 2. Paired studies of $PaCO_2$ (kPa) (mm Hg), Mean arterial blood pressure (MABP) (kPa) (mm Hg), CBF (ml/100 g/min), $CMRO_2$ (ml O_2/100 g/min), the ratio $CBF/CMRO_2$ and CVR (kPa 100 g/min/ml). The studies were performed during craniotomy for supratentorial cerebral tumors with the intravenous modification of the Kety and Schmidt method. The first study was performed about one hour after induction of anaesthesia (1), and the second study one hour later (2). The study indicates the effect of *unchanged anaesthetic level*. Mean values are indicated

Anaesthetic method	No.	$PaCO_2$ kPa (mm Hg)	MABP kPa (mm Hg)	CBF ml	$CMRO_2$ ml O_2	$CBF/CMRO_2$ ratio	CVR
Althesin 0.2 ml/kg/h N$_2$O 67%, fentanyl (1)	5	4.2 (32)	13.8 (104)	22	1.8	12.2	0.63
Althesin 0.2 ml/kg/h N$_2$O 67%, fentanyl (2)	5	3.6 (27)	12.2 (92)	20	1.9	10.5	0.61
Etomidate 0.5 µg/kg/h N$_2$O 67%, fentanyl (1)	7	4.0 (30)	12.3 (92)	25	2.3	10.6	0.49
Etomidate 0.5 µg/kg/h N$_2$O 56%, fentanyl (2)	7	3.9 (29)	12.5 (94)	22	2.2	10.0	0.57
Halothane 0.45% N$_2$O 67%, fentanyl (1)	7	4.0 (30)	11.5 (86)	35	2.7	13.0	0.33
Halothane 0.45% N$_2$O 67%, fentanyl (2)	7	4.0 (30)	10.3 (77)	34	2.6	13.1	0.30
Enflurane 1.0% N$_2$O 67%, fentanyl (1)	7	4.3 (32)	11.5 (86)	30	2.0	15.2	0.38
Enflurane 1.0% N$_2$O 67%, fentanyl (2)	7	4.2 (32)	11.1 (83)	27	1.9	14.5	0.41

Isoflurane 0.75% N_2O 67%, fentanyl (1)	7	4.1 (31)	9.7 (73)	31	2.1	14.2	0.31
Isoflurane 0.75% N_2O 67%, fentanyl (2)	7	3.9 (29)	8.8 (66)	29	2.0	15.5	0.30
Halothane 0.5%, N_2O 67%, fentanyl, pentobarbitone (1)	6	3.4 (26)	12.4 (93)	27	2.2	12.4	0.47
Halothane 0.5%, N_2O 67%, fentanyl, pentobarbitone (2)	6	3.4 (26)	12.3 (92)	29	2.3	12.4	0.43
Neurolept anaesth. Droperidol 0.2 mg/kg, fentanyl, N_2O (1)	10	4.5 (34)	12.9 (97)	30	2.3	13.0	0.43
Neurolept anaesth. Droperidol 0.2 mg/kg, fentanyl, N_2O (2)	10	4.3 (32)	12.0 (90)	28	2.5	11.2	0.43
Midazolam 0.125 mg/kg/h, N_2O, fentanyl (1)	10	4.3 (32)	11.0 (83)	28	2.7	10.4	0.39
Midazolam 0.125 mg/kg/h, N_2O, fentanyl (2)	10	4.2 (32)	10.9 (82)	23	2.4	9.6	0.47

* $P < 0.05$.

Table 3. Paired studies of $PaCO_2$ (kPa) (mm Hg), Mean arterial blood pressure (MABP) (kPa) (mm Hg), CBF (ml/100 g/min), $CMRO_2$ (ml O_2/100 g/min), the ratio CBF/$CMRO_2$ and CVR (kPa 100 g/min/ml). The studies were performed during craniotomy for supratentorial cerebral tumors with the intravenous modification of the Kety and Schmidt method. The first study was performed about one hour after induction of anaesthesia (1), and the second study one hour later (2). The study indicates the *effect of increased anaesthetic level.* Mean values are indicated

Anaesthetic method	No.	$PaCO_2$ kPa (mm Hg)	MABP kPa (mm Hg)	CBF ml	$CMRO_2$ ml	CBF/$CMRO_2$ ratio	CVR
Althesin 0.2 ml/kg/h N₂O 67%, fentanyl (1)	4	4.0 (30)	12.8 (96)	27	1.9	14.2	0.47
Althesin 0.5 ml/kg/h N₂O 67%, fentanyl (2)	4	3.8 (29)	11.5 (86)	20	1.5	13.3	0.58
Etomidate 0.5 µg/kg/h N₂O 67%, fentanyl (1)	7	4.3 (32)	12.8 (96)	35	2.5	13.7	0.37
Etomidate 1.0 µg/kg/h N₂O 67%, fentanyl (2)	7	4.4 (33)	12.2 (92)	23*	1.8*	13.1	0.53*
Halothane 0.45% N₂O 67%, fentanyl (1)	7	4.1 (31)	11.7 (88)	32	2.3	13.9	0.37
Halothane 0.9% N₂O 67%, fentanyl (2)	7	4.0 (30)	9.1 (68)*	36*	2.0*	18.0	0.25*

	n						
Enflurane 1.0% N$_2$O 67%, fentanyl (1)	6	4.3 (32)	11.4 (86)	28	1.9	15.0	0.41
Enflurane 2.0% N$_2$O 67%, fentanyl (2)	6	3.9 (29)*	8.8 (66)*	27	1.6*	16.9	0.33*
Isoflurane 0.75% N$_2$O 67%, fentanyl (1)	7	4.1 (31)	10.9 (82)	35	2.4	14.6	0.31
Isoflurane 1.5% N$_2$O 67%, fentanyl (2)	7	4.0 (30)	8.9 (67)*	33	1.9*	17.4	0.27
Halothane 0.5%, N$_2$O 67%, fentanyl, pentobarbitone (1)	10	3.6 (27)	11.0 (83)	24	1.8	13.2	0.47
Halothane 1.0%, N$_2$O 67%, fentanyl, pentobarb, and thiop. (2)	10	3.5 (26)	9.9 (74)*	20*	1.6*	12.6	0.50
Midazolam 0.125 mg/kg/h, N$_2$O 67%, fentan. (1)	10	4.0 (30)	9.5 (71)	23	2.1	11.0	0.41
Midazolam 0.25 mg/kg/h, N$_2$O 67%, fentanyl (2)	10	3.7 (28)	10.2 (77)	28	1.7	16.5	0.36

* $P < 0.05$.

Study 3 (Halothane 0.5% Supplemented with Thiopentone)

Patients: Sixteen patients (mean age 50 years, range 21–69 years) were studied. At CT scanning all patients had midline shifts less than 10 mm and the tumors were all localized supratentorially. Before the study all patients were awake, and undergoing treatment with steroids.

Anaesthesia: All patients were premedicated with pentobarbitone 1–2 mg/kg i.m. one hour before the induction of anaesthesia. For induction pentobarbitone 4 mg/kg, fentanyl 0.1 mg and pancuronium 0.1 mg/kg were given. After tracheal intubation, the anaesthesia was maintained with halothane 0.5% in nitrous oxide 67%. Additional fentanyl 0.1 mg was given at the time of skin incision and pancuronium was given in doses of 1–2 mg when necessary to provide complete relaxation. In six patients CBF was measured twice under these circumstances, the first measurement being performed before incision, and the second one hour later after opening of the dura. In five patients, thiopentone 8 mg/kg i.v. was supplemented during opening of the dura and the second CBF measurement was performed. In another five patients, thiopentone 16 mg/kg was administered i.v. before the second CBF measurements.

Results: In the group of six patients not subjected to loads of thiopentone, CBF and $CMRO_2$ averaged 27 ml/100 g/min and 2.2 ml O_2/100 g/min during the first flow measurement; and 29 ml/100 g/min and 2.3 ml O_2/100 g/min during the second flow run (Table 2).

In the group of 10 patients subjected to thiopentone loads after opening of the dura, the results of the average values of CBF and $CMRO_2$ were identical whether thiopentone loads of 8 or 16 mg/kg were applied. During the first measurement, CBF and $CMRO_2$ averaged 24 ml/100 g/min and 1.8 ml O_2/100 g/min respectively; and during the second flow run the values obtained were 20 ml/100 g/min for CBF ($P < 0.05$), and 1.6 ml O_2/100 g/min for $CMRO_2$ ($P < 0.05$) (Table 3).

Study 4 (Neurolept Anaesthesia)

Patients: Ten patients (mean age 57 years, range 39–68 years) with supratentorial cerebral tumors were included in the study. By CT scanning the midline shifts were found to be less than 10 mm in 7 patients, 15 mm in one patient and 20 mm in two patients. Preoperatively all patients were awake and receiving steroids.

Anaesthesia: The patients were premedicated with diazepam 10–15 mg perorally one hour before anaesthesia. Thiopentone 4–6 mg/kg, droperidol 0.2 mg/kg, fentanyl 5 μg/kg, pancuronium 0.15 mg/kg, and lidocaine

1.5 mg/kg were used for induction. After intubation the anaesthesia was maintained with nitrous oxide 67% in oxygen and fentanyl 4 μg/kg/h. CBF was measured twice, the first measurement being performed before incision and the second after opening of the dura.

Results: During the first flow measurement CBF and $CMRO_2$ averaged 30 ml/100 g/min and 2.3 ml O_2/100 g/min respectively. During the second flow run, CBF averaged 28 ml/100 g/min and $CMRO_2$ 2.5 ml O_2/100 g/min (Table 2).

Study 5 (Continuous Midazolam)

Patients: Twenty patients with supratentorial cerebral tumors were included (mean age 56, range 34–67 years). All patients were awake before anaesthesia and under treatment with steroids. At CT scanning the shifts of midline structures were below 15 mm in all patients.

Anaesthesia: The patients were premedicated with diazepam 10–20 mg perorally one hour before induction of anaesthesia. The anaesthesia was induced with midazolam 0.3 mg/kg, fentanyl 4 μg/kg, pancuronium 0.15 mg/kg and lidocaine 1.5 mg/kg. After intubation the anaesthesia was maintained with midazolam 0.4 mg/kg/h in 15 min, followed by reduction of infusion rate to 0.125 mg/kg/h; furthermore, nitrous oxide 67% in oxygen and supplements of fentanyl 3 μg/kg/h were given. In ten patients the infusion rate of midazolam was maintained during the operation and CBF was measured twice. In another ten patients, the infusion rate of midazolam was increased after the first CBF measurement and maintained at 0.25 mg/kg/h. With this infusion rate the second flow measurement was repeated.

Results: During unchanged midazolam infusion of 0.125 mg/kg/h, CBF and $CMRO_2$ averaged 28 ml/100 g/min, and 2.7 ml O_2/100 g/min during the first CBF measurement and 23 ml/100 g/min and 2.4 ml O_2/100 g/min during the second flow (Table 2).

In the other group where the midazolam infusion rate was increased from 0.125 to 0.25 mg/kg/h between the first and second CBF measurement, CBF and $CMRO_2$ averaged 23 ml/100 g/min and $CMRO_2$ 2.1 ml O_2/100 g/min, and during the second measurement CBF averaged 28 ml/100 g/min and 1.7 ml O_2/100 g/min (Table 3).

Study 6 (Halothane 0.45 and 0.9%)

Patients: 14 patients, mean age 63 years (range 18–75), with supratentorial cerebral tumours were included. By CT scanning the midline shift was found to be less than 15 mm in all patients. All patients were alert before operation, and under treatment with steroids.

Anaesthesia: One hour before anaesthesia the patients were pre-medicated with pentobarbitone 100–150 mg i.m. Thiopentone 4–6 mg/kg, fentanyl 0.1 mg, lidocaine 1.5 mg/kg and pancuronium 0.15 mg/kg were used for induction of anaesthesia. After intubation the anaesthesia was maintained with halothane 0.45% supplemented with nitrous oxide 67% in 7 patients and CBF was measured twice. In seven patients, the concentration of halothane was increased to 0.9% after the first CBF measurement and CBF was restudied.

Results: During unchanged halothane concentration of 0.45%, CBF and $CMRO_2$ averaged 35 ml/100 g/min and 2.7 ml O_2/100 g/min during the first CBF run, and 34 ml/100 g/min and 2.6 ml O_2/100 g/min during the second run (Table 2).

In the group of seven patients where the halothane concentration was increased from 0.45% to 0.9% between the first and the second measurement, the average value of CBF increased from 32 to 36 ml/100 g/min ($P < 0.05$), and $CMRO_2$ decreased from 2.3 to 2.0 ml O_2/100 g/min ($P < 0.05$) (Table 3).

Study 7 (Enflurane 1.0 and 2.0%)

Patients: Fourteen patients, mean age 60 years (range 43–70) with supratentorial cerebral tumours were included. Only patients with midline shift at CT scanning < 10 mm were included in the study. All patients were alert before operation and under treatment with steroids.

Anaesthesia: The patients were premedicated with pentobarbitone 100–150 mg i.m. one hour before anaesthesia. For Induction of anaesthesia, thiopentone 5–7 mg/kg, fentanyl 0.1 mg, lidocaine 1.5 mg/kg and pancuronium 0.15 mg/kg were used. After tracheal intubation the anaesthesia was maintained with enflurane 1.0% supplemented with nitrous oxide 67%, and CBF was measured twice in seven patients. In another seven patients, the enflurane concentration was increased to 2% after the first CBF measurement, and CBF was restudied with this concentration.

Results: With unchanged enflurane concentration of 1.0%, CBF and $CMRO_2$ averaged 30 ml/100 g/min and 2.0 ml O_2/100 g/min during the first study and 27 ml/100 g/min and 1.9 ml O_2/100 g/min during the second study (Table 2).

In the other group CBF and $CMRO_2$ averaged 45 ml/100 g/min and 1.9 ml O_2/100 g/min with the enflurane concentration of 1.0%. After increase in the enflurane concentration to 2.0% the average values of CBF and $CMRO_2$ were 36 ml/100 g/min and 1.6 ml O_2/100 g/min (P < 0.05) (Table 3).

Study 8 (Isoflurane 0.75 and 1.5%)

Patients: Fourteen patients with supratentorial cerebral tumours, mean age 57 years (range 22–75 years) were included. Preoperatively, all patients were awake, and under treatment with steroids. At CT scanning the midline shift was < 15 mm in all patients.

Anaesthesia: The patients were premedicated with pentobarbitone 100–150 mg i.m. one hour before operation. For induction of anaesthesia thiopentone 5–7 mg/kg, fentanyl 0.1 mg, pancuronium 0.15 mg/kg and lidocaine 1.5 mg/kg were given. After tracheal intubation the anaesthesia was maintained with isoflurane 0.75% supplemented with nitrous oxide 67% throughout the anaesthesia in 7 patients and CBF was measured twice. In another seven patients, the concentration of isoflurane was increased to 1.5% after the first CBF measurement, and CBF was restudied.

Results: In the seven patients with unchanged isoflurane concentration, CBF averaged 31 ml and 29 ml/100 g/min during the first and second flow study; and $CMRO_2$ averaged 2.1 and 2.0 ml O_2/100 g/min (Table 2).

In the other seven patients where the isoflurane concentration was increased from 0.75 to 1.5% between the first and the second study, CBF was unchanged (35 and 33 ml/100 g/min). $CMRO_2$ was 2.4 ml and 1.9 ml O_2/100 g/min (P < 0.05) (Table 3).

Discussion and Summary of the Results

In the studies 1–5 the method used for the measurements of CBF and $CMRO_2$ were identical and all the studies were performed during craniotomy for cerebral tumors. These principles were identical with recent studies using the inhalation agents, (halothane, enflurane and isoflurane), used in equipotent doses and supplemented with nitrous oxide 66% and fentanyl (Madsen *et al.* 1986, Madsen *et al.* 1987a, Madsen *et al.* 1987b). The results are shown in tables 2 and 3. The time of the CBF measurements in relation to induction of anaesthesia was practically the same in all studies; thus, the first CBF was measured about one hour after induction, before surgical incision; the second flow was measured one hour later, after opening of the dura and during surgical removal of the tumors. Moreover, all patients were suffering from supratentorial tumors, were awake before induction of anaesthesia and at CT scanning the majority of patients in each group had midline shifts below 10 mm. With the exception of study 6, where the average levels of $PaCO_2$ ranged from 3.4 to 3.6 kPa (26–27 mm Hg), mean $PaCO_2$ in the other studies were fairly identical, ranging between 4.0–4.5 kPa (30–34 mm Hg) during the first flow measurement, and 3.6–4.3 kPa (27–32 mm Hg) during the second, signifying that the degree of hyperventilation applied was identical in the different studies.

As regards premedication, pentobarbitone was used in study 1–3 and during the studies of halothane, enflurane and isoflurane, while diazepam was used in the studies of neurolept anaesthesia and midazolam. Theoretically, pentobarbitone in doses of 1–2 mg/kg might influence cerebral circulation and metabolism even two hours after administration, but not three hours.

As indicated, a dose related decrease in $CMRO_2$ was observed during althesin, etomidate, halothane, enflurane, isoflurane and midazolam anaesthesia. A dose dependent decrease in CBF was noticed during althesin and etomidate anaesthesia, while a tendency of CBF to increase was observed during anaesthesia with halothane but not with enflurane and isoflurane.

The ratio $CBF/CMRO_2$ was low (between 13 and 15) during althesin, etomidate, midazolam, halothane, isoflurane anaesthesia and neurolept anaesthesia and remained low during increasing doses of intravenous anaesthesia with althesin and etomidate but increased from 14.6 to 17.4 during isoflurane concentrations of 1.5%. During enflurane 1%, the ratio averaged 17.7, and remained unchanged during 2%.

CVR was high during althesin, etomidate, midazolam, neurolept anaesthesia and anaesthesia with halothane and pentobarbitone induction, and increased dose-dependently with althesin, etomidate and during thiopentone loads as a supplement to halothane anaesthesia. However, the CVR was considerably lower during anaesthesia with inhalation anaesthetics with thiopentone induction, and decreased dose-dependently when increasing doses of halothane, enflurane and isoflurane were applied.

As indicated, during IsoMAC anaesthesia with halothane, enflurane and isoflurane, CBF and CVR were practically identical and differed from the values obtained during intravenous anaesthesia with althesin, etomidate, and to some extent neurolept anaesthesia, where a lower CBF, and a higher CVR were obtained. A certain difference between halothane where either thiopentone or pentobarbitone were used for induction was observed, indicated by a lower value of CBF and $CMRO_2$ in the group where pentobarbitone was used for induction and a higher level of CVR in the same group. Furthermore, thiopentone loads of 8 mg/kg given during maintenance of halothane anaesthesia decreased CBF and $CMRO_2$, while an increase in the halothane concentration from 0.45 to 0.9% was followed by an increase in CBF, but a decrease in $CMRO_2$.

The results of the present studies are in accordance with experimental and clinical studies where the intravenous hypnotics (althesin, etomidate, midazolam and the drugs used in neurolept anaesthesia) or inhalation anaesthetics (halothane, enflurane and isoflurane) were administered without nitrous oxide. They confirm the findings of the few studies performed during craniotomy. Thus Eintrei *et al.* (1985) using local application of 133-Xenon for studying rCBF during craniotomy found an increase of 34% in rCBF during enflurane 1.1% and an increase of 168% during halothane 0.6% when administered during nitrous oxide anaesthesia. On the other hand isoflurane in 1% did not increase rCBF. The quantitative difference in results is supposed to be caused by the difference in methodology (regional contra global CBF) and by differences in concentrations of inhalation agents. Other studies of isoflurane used for controlled hypotension during craniotomy for cerebral aneurysm suggest unchanged CBF and a decrease in $CMRO_2$ at hypotension levels of 50–60 mm Hg, equivalent to administration of 1.5–2.5% isoflurane (Newmann *et al.* 1986, Roth *et al.* 1986, Madsen *et al.* 1987b).

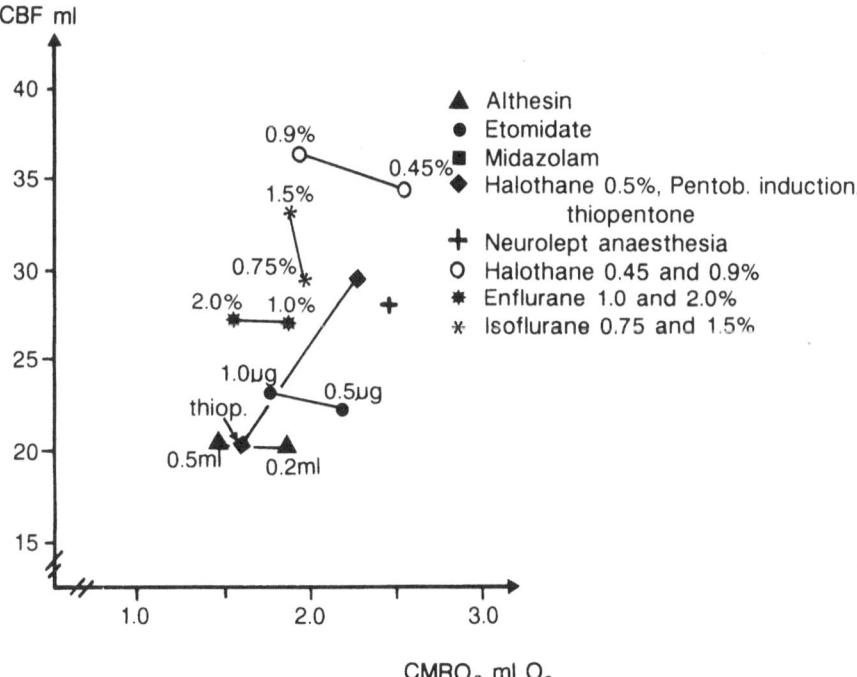

Fig. 10. Correlation between $CMRO_2$ (ml O_2/100 g/min) and CBF (ml/100 g/min) in paired studies of neuroanaesthesia in patients subjected to craniotomy for supratentorial cerebral tumors. The anaesthetic methods and concentration of anaesthetics are indicated

The relationship between $CMRO_2$ and CBF for the eight studies are shown in Fig. 10. Only data from the second studies measured during opening of the dura are included. The corresponding relationship between CVR and the ratio $CBF/CMRO_2$ is shown in Fig. 11. In the figures the inhalation agents halothane, enflurane and isoflurane and the intravenous anaesthetics althesin, etomidate neurolept anaesthesia and midazolam are indicated separately. Anaesthesia with pentobarbitone induction and halothane 0.5% in nitrous oxide follows the intravenous anaesthetics.

These observations suggest that from a cerebral hemodynamic point of view, the intravenous anaesthetic agents are preferable to inhalation anaesthetics as CBF, $CMRO_2$, $CBF/CMRO_2$ ratio generally are lower, and CVR higher, suggesting that at the same level of pressure of CPP and $PaCO_2$, the CBV is lower during intravenous anaesthesia, the surgical access and procedure therefore improved and the risks of intracranial hypertension reduced.

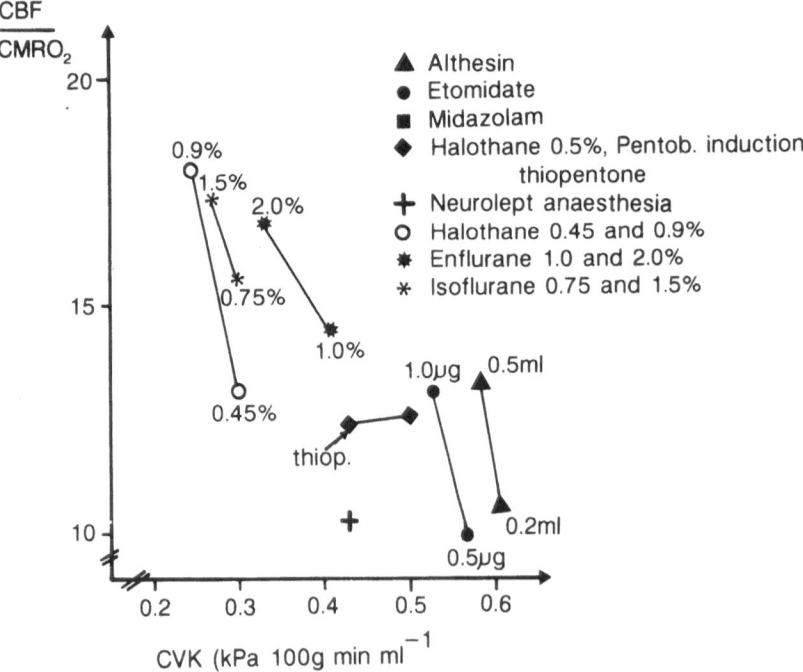

Fig. 11. The correlation between cerebral vascular resistance (CVR) ($kPa \times 100\,g \times min/ml$) and the ratio $CBF/CMRO_2$ in paired studies on neuro-anaesthesia in patients subjected to craniotomy for supratentorial cerebral tumors. The anaesthetic methods and concentration of anaesthetics are indicated

Dynamic Studies

Studies of CBF and metabolism during craniotomy presume stabilized haemodynamic conditions for 15–30 min, which at low flow rates represents the time necessary for calculation of CBF from the arterial and venous washout curves of 133 Xenon.

Initial slope index, calculated from the slope of the semilogarithmically displayed clearance curves after an arterial or venous bolus injection, or after inhalation of Xenon for 3 min, offers an alternative but has not been used during craniotomy in humans because the method assumes the position of detectors within the operative field.

Measurements of arterio-venous oxygen content difference ($AVDO_2$) have been used to evaluate the dynamic changes in CBF and $CMRO_2$. If the brain is perfused sufficiently in all regions to cover metabolic needs, $AVDO_2$ is equivalent to $CMRO_2/CBF$, and $1/AVDO_2 = CBF/CMRO_2$. At constant $CMRO_2$ levels, $1/AVDO_2$ is directly proportional to CBF.

During clinical anaesthesia for craniotomy six periods are of interest: The period of intubation, the period between induction of anaesthesia and exposure of the dura, with special reference to the skin incision and opening of the dura, the period of weaning of the hyperventilation and lastly, the recovery period after extubation.

In a recent study, $AVDO_2$ was repeatedly measured pre- and postoperatively during either halothane 0.5%, N_2O, fentanyl anaesthesia or neurolept anaesthesia. The MABP and $PaCO_2$ were found to be at the same levels during and after the two anaesthetic procedures. However, the levels of $AVDO_2$ were found to be higher during neurolept anaesthesia compared with halothane anaesthesia, signifying a relative state of luxury-perfusion during halothane anaesthesia. Furthermore, a significant decrease in $AVDO_2$, associated with an increase in MABP were found in both groups during incision, and after extubation. See figure 1 (Engberg *et al.* 1989).

In the period between intubation and opening of the dura, clinical studies of CBF and metabolism during neurolept anaesthesia and during continuous etomidate infusion have shown a fairly good correlation between $AVDO_2$ and CBF, and between jugular venous oxygen saturation and CBF (Cold *et al.* 1985, Cold *et al.* 1988). Thus, monitoring of $AVDO_2$ or jugular venous oxygen saturation can give an estimate of CBF before and after the autoregulatory stress of incision and is expected to be a guide to therapeutic intervention. Fig. 12 shows the correlation between jugular venous oxygen saturation and CBF during neurolept anaesthesia, suggesting that at saturations above 60% the increase in CBF might be precipitous and indicate therapeutic intervention.

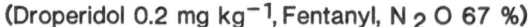

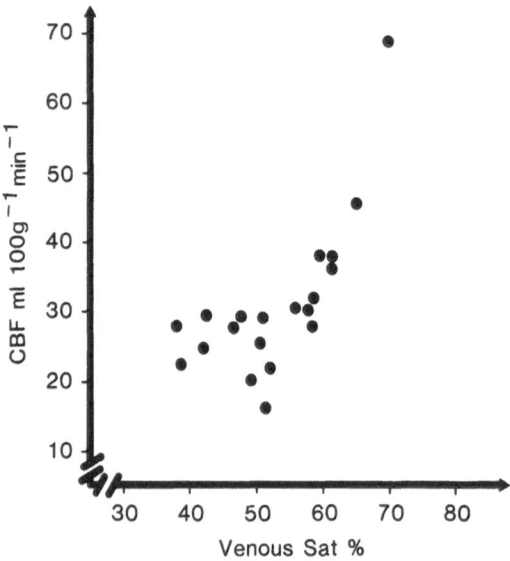

Fig. 12. The relationship between jugular venous oxygen saturation and CBF in 20 paired data from 10 patients subjected to craniotomy for supratentorial cerebral tumors in neurolept anaesthesia (Acta Anaesthesiol Scand 1988: 32: 310–315, with permission)

In patients subjected to craniotomy, the MABP increase observed during incision is blocked by scalp infiltration by plain bupivacaine (Rung et al. 1987), and studies of $AVDO_2$ during incision in patients subjected to craniotomy for cerebral tumors, have shown that the decrements in $AVDO_2$ can be blocked by scalp infiltration as well, suggesting that cerebral autoregulation is generally impaired in these patients and that CBF is therefore pressure dependent. The changes in $AVDO_2$ in patients with and without scalp infiltration are shown in Fig. 13. In some of the patients the decrease in $AVDO_2$ was equivalent to a CBF increase of about 100%, and this luxury perfusion was found to be of about 10–15 min duration, depending on the changes in MABP elicited by the incision (Engberg et al. 1990).

The decrease in $AVDO_2$ following extubation has been found to be of hours duration, and primarily dependent on the increase in MABP, while the degree of preoperative hyperventilation, changes in $PaCO_2$ after extubation, duration of operation, histological diagnosis of the tumor, the tumor size and localization and the extension of the operative exploration play minor parts (Asmussen et al. 1989). In Fig. 14 the changes in $AVDO_2$ during the postoperative period are correlated to changes in MABP.

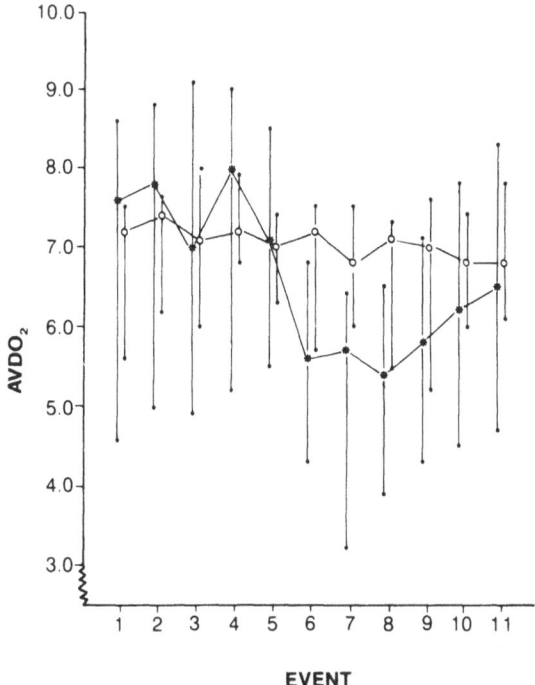

Fig. 13. The changes in arterio-venous oxygen content difference (AVDO$_2$) with time during scalp incision in 20 patients subjected to craniotomy in thiopentone induction, halothane 0.5%, nitrous oxide 67%, fentanyl. The patients were allocated into two groups. Group 1 (open circles) received plain bupivacaine 0.25% in the scalp 5 min before incision. Group 2 (star) received scalp infiltration of saline. AVDO$_2$ were measured at the following times. 1) Before infiltration, 2) 2 min after infiltration, 3) 5 min after infiltration, 4) 1 min before incision, 5) start of incision, 6) 2 min, 7) 4 min, 8) 6 min, 9) 10 min, 10) 15 min and 11) 20 min after incision. As indicated the AVDO$_2$ decreased in the saline group (Group 1), but remained constant in group 2. In both groups incision was followed by a significant increase in mean arterial blood pressure (MABP); however, the increase in MABP was significantly higher in group 2 (Acta Anaesthesiol Scand 1989, with permission)

Preliminary studies indicate that the beta adrenergic blocker esmolol effectively controls hypertension following intracranial surgery (Muzzi *et al.* 1988).

Future Clinical Studies of Cerebral Circulation and Metabolism During Anaesthesia

Until now human studies of rCBF and regional oxygen consumption during anaesthesia are few and only include the use of external detectors

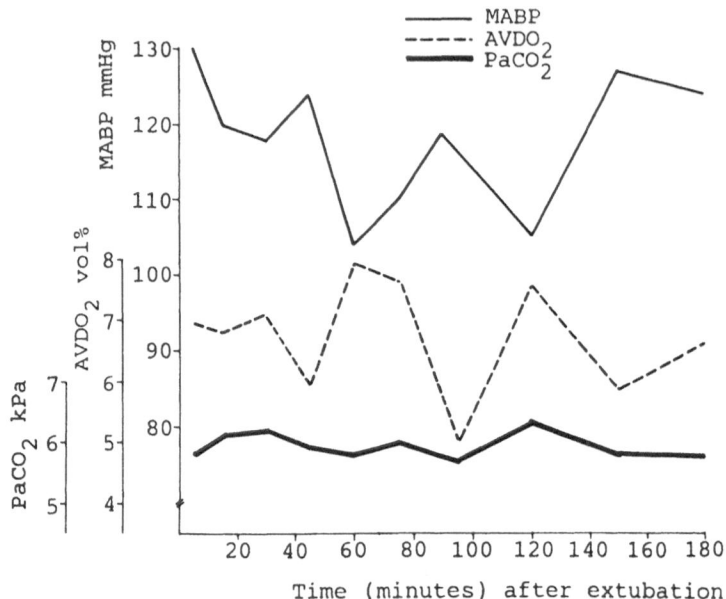

Fig. 14. Repeated postoperative studies of the arterio-venous oxygen content difference (AVDO₂) correlated to mean arterial blood pressure (MABP) (mm Hg), and the level of PaCO₂ (kPa) in a 48 year old patient subjected to craniotomy for a cerebral tumor (Acta Neurochir (Wien): 1989: 101: 9–17 with permission)

placed over the cerebral hemispheres or measurement of global CBF. Thus, no information is available on the effects of anaesthetics on deep cerebral structures. With the availability of SPECT, enhanced stable xenon CT scannings and PET, valuable information might be possible both in normal brains and when focal cerebral lesions are present. MR scans might further elucidate the effect of anaesthetics on anatomical structures when a mass expanding lesion is present. Experimental studies of the difference in effect of anaesthetic agents upon rCBF, regional glucose utilization, biochemistry of cerebral tissue and cerebrospinal fluid in normal brains and in brains subjected to incomplete and complete global or focal ischaemia are of importance for better understanding of the dynamic changes in cerebral circulation and metabolism effected by anaesthetics and a better and more rational choice of anaesthetic practice.

9. The Sitting Position

From a surgical aspect, the sitting position improves access to infratentorial structures and improves venous drainage. On the other hand the hydrostatic difference between cerebral veins and the right atrium augment risks of air embolism and the change in position might provoke postural hypotension.

In this review, two aspects of the sitting position shall be discussed: the influence on systemic blood pressure and consequently CPP, where hypotension might provoke a dangerous decrease in CPP and threatening cerebral oxygenation and the occurrence of air embolism with special reference to its prevention, diagnosis and treatment.

Cardiovascular Effects of the Sitting Position

The effects of change in position from supine to the sitting position on the cardiovascular performance have been studied repeatedly (Ward *et al.* 1966, Albin *et al.* 1974, Dalrymple *et al.* 1979). In healthy fully conscious volunteers, the stroke volume and cardiac output will decrease and the heart rate and systemic vascular resistance will increase, leaving MABP unchanged or increased. Marshall *et al.* (1983) compared the cardiovascular effects of four different anaesthetic techniques and found that morphine-nitrous oxide anaesthesia compared with enflurane, halothane and neurolept anaesthesia resulted in the least impairment when positioning was changed from supine to the sitting position. Tindall *et al.* (1967) found that hypocapnia in the sitting position decreased carotid artery flow by 48% and increased CVR. During anaesthesia the decrease in MABP is not always apparent (Tausk and Miller 1983). In a study by Albin *et al.* (1976), 32% of the patients had a 10–20% drop in blood pressure in reaching the sitting position, while 11% became temporarily hypertensive. In other studies, the incidence of hypotension is between 5 and 10% (Millar 1972, Young *et al.* 1986).

Intravenous infusion of crystalloids before change in position from supine to the sitting position has been recommended. Matjasko *et al.* (1985) recommended doses between 7–14 ml/kg. Nevertheless vasopressors are often necessary to prevent or treat postural hypotension (Millar 1972, Tinker and Vandam 1972). The use of bandages on the legs to promote venous return might to some degree prevent hypotension (Tinker and

Vandam 1972, Dalrymple *et al.* 1979). However Geevarghese (1977) has observed that this precaution does not always prevent postural hypotension and other studies indicate the importance of abdominal compression to protect against gravitational acceleration (Wood and Lambert 1952, Burton and Krutz 1975). Thus, the antigravity suit has been recommended by several authors (Burton and Krutz 1975, Geevarghese 1977). With an inflation pressure of 25 mm Hg in the antigravity suit, the necessary volume of intravenous fluid and the use of vasopressors were less in patients supplied with an antigravity suit, compared with patients only supplied with elastic leg bandages (Freuchen 1959). In a recent study of the antigravity suit (G suit, aviation type) inflation of the suit to 3.3 kPa (25 mm Hg) in the sitting position was followed by a significant increase in systolic blood pressure and CVP. The addition of 5 mm Hg PEEP during suit inflation was accompanied by a further increase in CVP (Brodrick and Ingram 1988).

Cerebrovascular and Spinal Cord Dynamics

The effect of the sitting position on cerebrovascular and spinal dynamics have been studied experimentally. In dogs subjected to intracranial hypertension by inflation of an epidural balloon a significant decrease in cerebral and spinal CBF occurs in the sitting position, while $CMRO_2$ remains fairly constant (Ernst *et al.* 1988). In a preliminary study Jensen *et al.* (1989) did not find any decrease in CBF when placing the patient in the sitting position.

Air Embolism

A pressure gradient of 5 cm between the upper pole of a wound and the right atrium is sufficient to allow air to pass into the venous system (Albin *et al.* 1983). Patients in the sitting position, where a gradient of 30 cm is usual, are therefore especially vulnerable to air embolism. The clinical effects of air embolism are determined by several factors. Butler and Hills (1979) state that a large volume of air may exceed the lung's capacity to dissipate the air. Experimental studies indicate that the rate of air entry is related to the morbidity (Wollfe and Robertson 1935, Richardson *et al.* 1937), and studies in dogs subjected to continuous air infusion, indicate a progressive increase in CVP, an abrupt increase in pulmonary arterial pressure, a progressive decrease in peripheral resistance and a compensatory increase in cardiac output; blood pressure decreased moderately until compensation was exceeded, at which point blood pressure decreased sharply. The electrocardiographic changes involved peaking of the p waves and later depression of the ST segment. Changes in heart sounds occurred only when significant cardiovascular decompensation had already occurred (Adornato *et al.* 1978). The same authors studied the effects of a bolus injection of air and found an increase in CVP but a decrease in pulmonary arterial pressure, ST depression on ECG and shock. The threshold for the lung to filter air is 0.30 ml/kg/min (Butler and Hills 1979, Butler and Hills 1985). This threshold is reduced during halothane and isoflurane anaesthesia (Katz *et al.* 1985) but increased during nitrous oxide anaesthesia and PEEP (Butler *et al.* 1985, Butler *et al.* 1986). In dogs subjected to halothane anaesthesia, the threshold of transpulmonary passage of venous air was 0.04 ml/kg/5 sec, but about 1 ml/kg/5 sec during pentobarbitone anaesthesia with or without halothane (Yahagi and Furuya 1987).

The fate of air emboli in the pulmonary circulation has been studied in dogs with a thoracic window. The findings indicate that venous air emboli are eliminated primarily in the pulmonary arterioles, as a result of molecular diffusion of gas across the arteriolar wall into alveolar spaces (Presson *et al.* 1989).

Episodes of air embolism occur predominantly early during the operative procedure (Albin *et al.* 1976, Standefer *et al.* 1984), especially during stripping of muscles and fascia (Leivers *et al.* 1971), drilling of the skull (Ericsson 1964) and during fascial layer closure (O'Higgins 1970).

The occurrence of air embolism in the sitting position might be fatal, partly because the air will obstruct venous return to the heart and consequently reduce blood pressure and provoke hypoxia and partly because the embolus might pass into the general circulation, either through the lungs or through a patent foramen ovale. In a recent study, systemic air embolism occurred in about 2% of all patients having venous air embolism, and in 0.5% of all patients subjected to the sitting position (Matjasko *et al.* 1985). Coronary air embolism might present itself with ST elevation or depression or widening of the ORS complex, and after cerebral artery air embolism epileptogenic activity has been observed (Clayton *et al.* 1985), and postoperatively, cerebral deterioration as a result of cerebral ischaemia might occur (Gronert *et al.* 1979). At postmortem examination a patent foramen ovale is present in about 20% of the population (Edwards 1968). Preoperative contrast echocardiography provides a safe method of detection of right-to-left shunts (Higgins *et al.* 1984) and using this method Guggiare *et al.* (1985) found an incidence of 13% of right-to-left shunts and it has been argued that on a patient in whom a right-to-left shunt has been detected, surgery in the sitting position should be avoided (Brodrick 1987). The occurrence of right-to-left embolization of air is dependent on a right-to-left pressure gradient over the heart. Studies in dogs have shown that venous air embolism causes an increase in pulmonary pressure and CVP and an increase in the right-to-left pressure gradient (Adornato *et al.* 1978, Perkins-Pearson *et al.* 1982).

Although PEEP has been used clinically in the prevention of air embolism (Voorhies *et al.* 1983), the effect of PEEP on the right-to-left pressure gradient is discussed. Thus, studies in sheep and dogs (Metha and Sokoll 1981, Pfitzer and McLean 1987), and man (Perkins and Bedford 1984) indicate an increase in the pressure gradient, when PEEP or active lung inflation for 10 sec is applied, while other studies indicate an unchanged pressure gradient, in the presence of PEEP (Pearl and Lawson 1986, Zasslow *et al.* 1986). Furthermore, PEEP raises the threshold of the lung for the filtration of air and it is consequently unlikely that PEEP will increase the incidence of paradoxical air embolism (Butler *et al.* 1986). In a recent study in dogs PEEP at 15 cm H_2O was compared with neck compression at 40 mm Hg. Neither maneuver affected CBF measured by microsphere technique or $CMRO_2$. Incanine neck vein compression elevates sagital sinus pressure above atmospheric pressure, but PEEP did not, suggesting that neck compression used as a prophylactic measure is superior to PEEP (Toung *et al.* 1988). Preliminary studies of jugular venous bulb pressure in patients subjected to neck compression and for PEEP in the sitting position show that neck compression elevates venous pressure to positive values, an effect not observed with PEEP upto 20 cm H_2O or during inflation of anti-G suit (Cold *et al.* 1990).

The incidence of air embolism in the sitting position depends on the methods used to detect embolization and the type of operation, suboccipital craniectomies being particularly susceptible (Matjasko *et al.* 1985, Standefer *et al.* 1984). Precordial Doppler devices introduced by Michenfelder *et al.* (1972) have been found to be sensitive to the quantity of air in the range of 0.12–0.24 ml/kg (English *et al.* 1978, Glenski *et al.* 1986). Even more sensitive is transoesophageal Doppler with a sensitivity of 0.05–0.2 ml/kg (Martin and Colley 1983) and transoesophageal echocardiography with a sensitivity of 0.05–0.19 ml/kg (Glenski *et al.* 1986). End tidal CO_2 analysis can detect air embolisms of 0.18–0.63 ml/kg (Edmonds-Seal *et al.* 1971, Glenski *et al.* 1986), while changes in CVP, ECG, MABP and oxygen tensions are detected at even greater air embolisms (English *et al.* 1978, Glenski *et al.* 1986). Recent studies in upright sheep of venous pressure in the jugular bulb suggest that an increase in pressure provides an earlier and more reliable warning than did a Doppler probe (Pfitzner *et al.* 1987). In a recent study of air bolus embolism in dogs, end-tidal nitrogen and pulmonary arterial pressure increased earlier than did change in end tidal CO_2; however, the changes were simultaneously recorded after large infusion doses (Matjasko et al. 1987).

The lateral sitting position has been found to have advantages over the conventional sitting position because it is possible to change to the lateral supine position without disturbing the surgical procedure (Garcia-Bengochea et al. 1976) and in a clinical study, the lateral sitting position with 45 degree elevation of the head and thorax, CPP and ICP was found to be more stable compared to the upright sitting position and the lateral position with 20 degree elevation of the head and thorax (Calliauw et al. 1987).

Mandatory to a successful resuscitation of patients with air embolism is the speed of diagnosis and treatment (Ericsson et al. 1964). Immediate information to and from the surgeon as soon as air embolization is diagnosed is important, to assess the point of entry of air (Michenfelder et al. 1969). Bimanual neck compression by hand has been recommended (Michenfelder et al. 1969, Tausk and Miller 1983), or inflation of an inflatable neck tourniquet or neck tie (Hewer and Logue 1962, Buckland and Manners 1976, Sale 1984). Immediate withdrawal of nitrous oxide and the administration of pure oxygen will control and reduce the expansion of air bubbles in the blood (Nunn 1959, Munson 1971). Aspiration through a central caval catheter or a Swan Ganz catheter may remove some of the air (Bedford et al. 1981), but this attempt is often disappointing because it is difficult to secure the ideal position of the catheter (Sink et al. 1976). However, a special pulmonary artery catheter introducer sheath has recently been studied experimentally. Air aspiration through this catheter is more effective compared with pulmonary catheters and improve survival (Bowdle and Artru 1988). Hypotension should be treated with a rapid infusion of crystalloids and colloids (Matjasko et al. 1985) and during persistence of hypotension vasopressors such as ephedrine, dopamine or angiotensin can be of help. If these measures are ineffective in restoring circulation, repositioning in the left lateral position might improve circulation by its gravity effect on the position of the air embolism in the heart (Hamby and Terry 1952, Marshall 1965, Alvaran et al. 1978).

Experiments with cats have demonstrated that pre-treatment with intravenous lidocaine reduces neural decrement and increases the recovery of neural function after acute cerebral ischaemia induced by air embolism (Evans et al. 1984). Other experimental studies have shown that pretreatment with perfluorocarbon, which increases solubilities of oxygen, nitrogen and carbon dioxide may potentially protect against the effects of air embolism (Spiess et al. 1986) and help to attenuate some of the detrimental cardiovascular effects of the air embolism (Tuman et al. 1986).

Conclusion

As stated by Standefer *et al.* 1984 and Matjasko *et al.* (1985), the sitting position is safe indicated by low morbidity and mortality. In a retrospective study of 579 patients undergoing the sitting or prone position for fossa posterior surgery, venous air embolism occurred more often in the sitting position (45% versus 12%). However, no morbidity or mortality was attributed to venous air embolism and the incidence of hypotension or preoperative cardiopulmonary complications were not increased in the sitting position suggesting that selection of a head-down surgical position solely to decrease peroperative complications is not justified (Black *et al.* 1988). Patients with ischaemic heart disease, severe hypertension and known right-to-left shunts should not be placed in the sitting position. A prophylactic regime including anti-G suit, intravenous crystalloids, PEEP combined with reasonable monitoring for detection of venous air embolism should be considered as securing a safe procedure.

10. General Summary

The effects of anaesthetics on cerebral circulation, metabolism, ICP, brain tissue metabolites are reviewed. A summary of clinical studies of CBF and $CMRO_2$ during craniotomy for cerebral tumors then follows as well as a review of the sitting position used in neurosurgical anaesthesia.

Traditionally, drugs used in anaesthesiological practice are divided into four groups including inhalation agents, hypnotics, central analgetics and muscle relaxants. The authors give a review of the literature concerning the effects of anaesthetics on CBF, $CMRO_2$, glucose utilization, ICP and electroencephalograms. Inhalation agents (halothane, enflurane and iso-flurane) give rise to a dose-dependent increase in CBF and a decrease in $CMRO_2$. During anaesthesia with these agents the CO_2 reactivity is generally intact, while the cerebral autoregulation is impaired; a normalization of the cerebral autoregulation however occurs during hypocapnia. Nitrous oxide elicits an increase in CBF whereas the effect on $CMRO_2$ is unpredictable. As a supplement to halothane, enflurane or isoflurane, and to some hypnotics, nitrous oxide provokes an increase in CBF and further impairment of the cerebral autoregulation. In contrast to the inhalation anaesthetics, hypnotic drugs (barbiturate, althesin, etomidate, propofol), diazepines (diazepam and midazolam), and fentiazin derivates like droperidol used in neurolept anaesthesia give rise to an associated decrease in CBF as well as $CMRO_2$. With hypnotics in high doses the electro-encephalogram becomes silent and $CMRO_2$ is not decreased with a further increase in dose. The maximum decrease in $CMRO_2$ by hypnotics is 50% during normothermia. Ketamine as the only drug gives rise to an increase in CBF as well as $CMRO_2$, associated with activation of the EEG. Given in small intravenous doses during normocapnia, central analgetics do not affect cerebral circulation or metabolism. However, in spontaneously breathing subjects, the depression of the respiratory center might give rise to hypercapnia, which provokes an increase in CBF and ICP. During spontaneous breathing central analgetics might therefore be dangerous in patients with reduced intracranial compliance or space-occupying lesions. The central analgetics fentanyl and especially alfentanil and sufentanil induce a fall in CBF and oxygen consumption. Non-depolarizing muscle relaxants do not affect CBF, ICP or metabolism, while succinylcholine provoke an increase in CBF, an increase in $PaCO_2$ and activation of the EEG.

Studies of CBF and $CMRO_2$ during craniotomy for cerebral tumors are reviewed. In patients with small supratentorial cerebral tumors, the authors review present studies of CBF and $CMRO_2$. Eight principles of neuro-anaesthesia have so far been investigated, including intravenous techniques with althesin, etomidate, neurolept anaesthesia and midazolam, and the studies have been supplemented by studies of the inhalation anaesthetics halothane, enflurane and isoflurane. In all studies the basic anaesthesia was supplied by nitrous oxide 67% and fentanyl and moderate hypocapnia was applied during the anaesthesia. An intravenous modification of the Kety and Schmidt technique using the 30 min arterial and jugular venous washout curves was applied for the calculation of CBF and $CMRO_2$ was calculated as the product of the arterio-venous oxygen content difference and CBF. The studies confirm the experimental and clinical studies: The intravenous hypnotics, benzodiazepines and droperidol give rise to dose-dependent decrease in CBF and $CMRO_2$ and an increase in cerebrovascular resistance (CVR). In the group of inhalation anaesthetics a dose-dependent decrease in $CMRO_2$ was observed, while a dose-dependent increase in CBF was only observed during halothane anaesthesia but not during enflurane and isoflurane anaesthesia. However, CVR and cerebral perfusion pressure (CPP) decreased dose-dependently during all inhalation anaesthetics and an increase in the ratio $CBF/CMRO_2$ was generally observed. In all studies the surgical approach was relatively uncomplicated and it was possible to control brain prolapse with intra-venous mannitol. In conclusion, these studies suggest that all eight anaesthetic procedures can be used safely as the surgery was uncomplicated and the tendency towards a brain prolapse after opening of the dura was of minor significance. However, the studies of CBF and $CMRO_2$ suggest that the intravenous techniques including althesin, etomidate, neurolept anaesthesia and midazolam anaesthesia are superior to inhalation agents because the suppression of cerebral circulation and metabolism is more pronounced.

The dynamic changes of cerebral circulation during craniotomy for cerebral tumors have been investigated by repeated studies of the $AVDO_2$. The authors refer especially to three periods: the period between induction of anaesthesia and opening of the dura, the period of surgical incision and the immediate postoperative period. In the period between induction of anaesthesia and exposure of dura, repeated paired studies of CBF, venous jugular saturation and $AVDO_2$ suggest a close correlation between venous jugular saturation and CBF during continuous etomidate anaesthesia and neurolept anaesthesia. Thus, studies of jugular venous saturation might predict global CBF and consequently be a guide to therapeutic intervention i.e. intensified hyperventilation or supplements with hypnotics. During the period of incision, a controlled study of scalp skin infiltration with plain bupivacain 0.25% suggests that the stress-induced increase in blood pressure can be prevented and furthermore, skin incision in patients with bupivacaine scalp infiltration is not followed by a fall in $AVDO_2$, which otherwise might occur as a result of the increase in blood pressure and abolished cerebral autoregulation. In the period after anaesthesia and extubation a decrease in $AVDO_2$ has been observed. It is argued that this fall, is generally a result of an increase in blood pressure during the first postanaesthetic period but other factors like the age of the patients, length of the operation and level of $PaCO_2$ might also play a part.

The effects of drugs used in controlled hypotension are reviewed. The literature concerning sodium nitroprusside, nitroglycerin, trimethapan, adenosine and the use of calcium blockers like esmerol and drugs with combined alfa and beta blocking properties are reviewed. With the exception of trimethapan and drugs with alfa and beta blocking properties, all drugs are cerebral vasodilators and experimental and clinical studies have unveiled an increase in CBF and ICP when these drugs are used for controlled hypotension. Adenosine seems to fulfil many of the demands asked for. It is non-toxic, the effect is controllable and it is the natural mediator of dilation during autoregulation processes and hypoxia. Furthermore, adenosine stabilizes the central hemodynamics and preserve organ blood flow. An alternative is isoflurane-induced controlled hypotension. Isoflurane preserves cardiac output and decreases peripheral resistance; the suppression of cerebral metabolism is supposed to be an advantage. Moreover, clinical studies indicate that the ischaemic threshold during isoflurane anaesthesia is about 10 ml/100 g/min, compared with 15 ml during enflurane and 20 ml during halothane anaesthesia. On the other hand experimental studies of focal complete or incomplete ischaemia have not disclosed a protective effect of isoflurane.

References

Abramson NS, Safar P, Detre K *et al* (1983) Results of a randomized clinical trial of brain resuscitation with thiopental. Anesthesiology 59: A101

Abou-Madi MN, Keszler H, Yacoub JM (1977) Cardiovascular reactions to laryngoscopy and tracheal intubation following small and large doses of lidocaine. Canad Anaesth Soc J 24: 12–19

— Trop D, Villemure JG (1983) Effect of changing $PaCO_2$ on intracranial pressure response to bolus infusion of mannitol. Anesthesiology 59: A391

Abou-Madi M, Trop D, Abou-Madi N, Ravussin P (1987) Does a bolus of mannitol initially aggravate intracranial hypertension? Br J Anaesth 59: 630–639

Adams RW, Gronert GA, Sundt TM, Michenfelder JD (1972) Halothane, hypocapnia, and cerebrospinal fluid pressure in neurosurgery. Anesthesiology 37: 510–517

— Cucchiara RF, Gronert GA, Messick JM, Michenfelder JD (1981) Isoflurane and cerebrospinal fluid pressure in neurosurgical patients. Anesthesiology 54: 97–99

Adornato DC, Gildenberg MD, Ferrario CM, Smart J, Frost EAM (1978) Pathophysiology of intravenous air embolism in dogs. Anesthesiology 49: 120–127

Aitken PG, Schiff SJ (1986) Barbiturate protection against hypoxic neuronal damage *in vitro*. J Neurosurg 65: 230–232

Albin MS, Jannetta PJ, Maroon JC, Tung A, Millen JE (1974) Anaesthesia in the sitting position. IV, Europ Congress Anaesth, Madrid. Excerpta Medica, Amsterdam, pp 775–778

— Babinski MF, Maroon JC, Jannetta PJ (1976) Anesthestic management of posterior fossa surgery in the sitting position. Acta Anaesth Scand 20: 117–128

— — Gilbert J, Smith SL (1983) Venous air embolism is not restricted to neurosurgery. Anesthesiology 59: 151

— (1984) The paradox of paradoxic air air embolism-PEEP, Valsalva and patent foramen ovale. Should the sitting position be abandoned. Anesthesiology 61: 222–223

— Bunegin L, Gelineau J (1986) ICP and CBF reactivity to isoflurane and nitrous oxide during normocapnia, hypocapnia and intracranial hypertension. In: Miller JD, Teasdale GM, Rowan JO, Galbraith SL, Mendelow AD (eds) Intracranial Pressure VI. Springer, Berlin Heidelberg, pp 719–724

— Bunegin L, Rasch J, Gelineau J, Ernst P (1989) Ketamine hydrochloride fails to protect against acute global hypoxia in the rat. Anesth Analg 68: S8

Albrecht RF, Miletich DJ, Rosenberg R, Zahed B (1977) Cerebral blood flow and metabolic changes from induction to onset of anesthesia with halothane or pentobarbital. Anesthesiology 47: 252–256

— — Madala LR (1983) Normalization of cerebral blood flow during prolonged halothane anesthesia. Anesthesiology 58: 26–31

Alexander SC, Wollman H, Cohen PJ, Chase PE, Behar M (1964) Cerebrovascular response to $PaCO_2$ during halothane anesthesia in man. J Appl Physiol 19: 561–565

— Smith TC, Strobel G, Stephen GW, Wollman H (1968) Cerebral carbohydrate metabolism of man during respiratory and metabolic alkalosis. J Appl Physiol 24: 66–72

Algotsson L, Messeter K, Nordström CH, Ryding E (1988) Cerebral blood flow and oxygen consumption during isoflurane and halothane anesthesia in man. Acta Anaesthesiol Scand 32: 15–20

Altenburg BM, Michenfelder JD, Theye RA (1969) Acute tolerance to thiopental in canine cerebral oxygen consumption studies. Anesthesiology 31: 443–448

Alvaran SB, Toung JK, Graff TE, Benson DW (1978) Venous air embolism: Comparative merits of external cardiac massage, intracardiac aspiration, and left lateral decubitus position. Anesth Analg 57: 166–170

Anderson RE, Sundt TM (1983) Brain pH in focal cerebral ischaemia and the protective effects of barbiturate anesthesia. J Cereb Blood Flow Metab 3: 493–497

Archer DP, Labrecque P, Tylor JL, Meyer E, Trop D (1987) Cerebral blood volume is increased in dogs during administration of nitrous oxide or isoflurane. Anesthesiology 67: 642–648

Arden JR, Holley FO, Stanski DR (1986) Increased sensitivity to etomidate in the elderly: Initial distribution versus altered brain response. Anesthesiology 65: 19–27

Arnér S, Gordon E (1976) The antagonist effect of naloxone hydrochloride after neuroleptanaesthesia during neurosurgery. Acta Anaesth Scand 20: 201–206

Arnfred I, Secher O (1962) Anoxia and barbiturates. Arch Int Pharmacodyn Ther 139: 67–74

Artru AA, Steen PA, Michenfelder JD (1980) Cerebral metabolic effects of naloxone administered with anesthetic and subanesthetic concentrations of halothane in the dog. Anesthesiology 52: 217–220

— Michenfelder JD (1981) Influence of hypothermia or hyperthermia alone or in combination with pentobarbital or phenotoin on survival time in hypoxic mice. Anesth Analg 60: 867–870

— (1983) Relationship between cerebral blood volume and CSF pressure during anesthesia with halothane or enflurane in dogs. Anesthesiology 58: 533–539

— (1984a) Isoflurane does not increase the rate of CSF production in the dog. Anesthesiology 60: 193–197

— (1984b) Relationship between cerebral blood volume and CSF pressure during anesthesia with isoflurane or fentanyl in dogs. Anesthesiology 60: 575–579

— (1984c) Effects of enflurane and isoflurane on resistance to reabsorption of cerebrospinal fluid in dogs. Anesthesiology 61: 529–533

— (1984d) Effects of halothane and fentanyl anesthesia on resistance to reabsorption of CSF. J Neurosurg 60: 252–256

Artru AA, Colley PS (1984) Cerebral blood flow responses to hypocapnia during hypotension. Stroke 15: 878–883

— (1986a) Cerebral metabolism and EEG during combination of hypocapnia and isoflurane-induced hypotension in dogs. Anesthesiology 65: 602–608

— (1986b) Cerebral metabolism and the electroencephalogram during hypocapnia plus hypotension induced by sodium nitroprusside or trimethaphan in dogs. Neurosurgery 18: 36–44

— Wright K, Colley PS (1986) Cerebral effects of hypocapnia plus nitroglycerin-induced hypotension in dogs. J Neurosurg 64: 924–931

— (1988a) Dose-related changes in the rate of cerebrospinal fluid formation and resistance to reabsorption of cerebrospinal fluid following administration of thiopental, midazolam, and etomidate in dogs. Anesthesiology 69:.541–546

— (1988b) Survival time during hypoxia: Effects of nitrous oxide, thiopental, and hypothermia (Editorial) Anesth Analg 67: 913–916

— Katz RA (1989) Cerebral blood volume and CSF pressure following administration of ketamine in dogs; modification by pre- or posttreatment with hypocapnia or diazepam. J Neurosurg Anesthesiol 1: 8–15

— (1989) Flumazenil reversal of midazolam in dogs: Dose-related changes in cerebral blood flow, metabolism, EEG, and CSF pressure. J Neurosurg Anesthesiol 1: 46–55

Ashton D, VanReempts J, Wauquier A (1981) Behavioral, electroencephalographic and histological study of the protective effect of etomidate against histotoxic dysoxia produced by cyanide. Arch Int Pharmacodyn Ther 254: 196–213

Askitopoulou H, Whitwam JG, Al-Khudhairi D, Chakrabarti M, Bower S, Hull CJ (1985) Acute tolerance to fentanyl during anesthesia in dogs. Anesthesiology 63: 255–261

Asmussen J, Cold GE, Elkjær S, Herlufsen P, Melsen NC, Engberg M, (1989) Changes in $AVDO_2$ in the postoperative period in patients subjected to craniotomy for supratentorial cerebral tumors. Acta Neurochir (Wien) 101: 9–17

Astrup J, Skovsted P, Gjerris F, Sørensen HR (1981a) Increase in extracellular potassium in the brain during circulatory arrest: Effects of hypothermia, lidocaine, and thiopental. Anesthesiology 55: 256–262

— Møller Sørensen P, Sørensen HR (1981b) Inhibition of cerebral oxygen and glucose consumption in the dog by hypothermia, pentobarbital, and lidocaine. Anesthesiology 55: 263–268

— (1982) Energy-requiring cell functions in the ischemic brain. Their critical supply and possible inhibition in protective therapy. J Neurosurg 56: 482–497

— Rosenørn J, Cold GE, Bendtsen A, Møller Sørensen P (1984) Minimum cerebral blood flow and metabolism during craniotomy. Effect of thiopental loading. Acta Anaesthesiol Scand 28: 478–481

Auer LM, Haselsberger K (1987) Effect of intravenous mannitol on cat pial arteries and veins during normal and elevated intracranial pressure. Neurosurgery 21: 142–146

Azar I, Turndorp H (1979) Severe hypertension and multiple atrial premature contractions following nalaxone administration. Anesth Analg 58: 524–525

Bachofen M (1988) Dämpfung des Blutdruckanstieges bei der Intubation: Lidocain oder Fentanyl? Anaesthesist 37: 156–161

Balslev-Jørgensen B, Misfeldt BB (1975) Intracranial pressure during recovery from nitrous oxide and halothane anesthesia in neurosurgical patients. Br J Anaesth 47: 977–981

Barker J, Harper AM, McDowall DG, Fitch W, Jennett WB (1968) Cerebral blood flow, cerebrospinal fluid pressure and EEG activity during neuroleptanalgesia induced with dehydrobenzperidol and phenoperidine. Br J Anaesth 40: 143–144

— (1987) Nitrous oxide in neurosurgical anaesthesia (editorial). Br J Anaesth 59: 146–147

Barry DI, Strandgaard S, Graham DI, Braedstrup O, Svendsen UG, Vorstrup S, Hemmingsen R, Bolvig TG (1982) Cerebral blood flow in rats with renal and spontaneous hypertension: Resetting of the lower limit of autoregulation. J Cereb Blood Flow Metab 2: 347–353

Baskin DS, Hosobuchi Y (1981) Nalaxone reversal of ischemic neurological deficits in man. Lancet ii: 272–275

Batjar HH, Frankfurt AI, Purdy PD, Smith SS, Samson DS (1988) Use of etomidate, temporary arterial occlusion, and intraoperative angiography in surgical treatment of large and giant cerebral aneurysms. J Neurosurg 68: 234–240

Baughman VL, Hoffman WE, Miletich DJ, Albrecht RF (1986) Effects of phenobarbital on cerebral blood flow and metabolism in young and aged rats. Anesthesiology 65: 500–505

— — Albrecht RF, Militich DJ (1987a) Cerebral vascular and metabolic effects of fentanyl and midazolam in young and aged rats. Anesthesiology 67: 314–319

— — Miletich DJ, Albrecht RF (1987b) Neurologic outcome following regional cerebral ischemia with methohexital, midazolam, and etomidate. Anesthesiology 67: A582

— — — — Thomas C (1988) Neurologic outcome in rats following incomplete cerebral ischemia during halothane, isoflurane, or N_2O. Anesthesiology 69: 192–198

— — — — (1989a) Cerebral metabolic depression and brain protection produced by midazolam and etomidate in the rat. Neurosurg Anesthesiol 1: 22–28

— — — — (1989b) Isoflurane vs. methohexital during incomplete cerebral ischaemia in the rat. Anesth Analg 68: S19

— — Thomas C, Albrecht RF, Miletich DJ (1989c) The interaction of nitrous oxide and isoflurane with incomplete cerebral ischemia in the rat. Anesthesiology 70: 767–774

Becker GL, Pelligrino DA, Miletich DJ, Albrecht RF (1986) The effect of nitrous oxide on oxygen consumption by isolated cerebral cortex mitochondria. Anesth Analg 65: 355–359

Bedford RF, Persing JA, Pobereskin L, Butler A (1980a) Lidocaine or thiopental for rapid control of intracranial hypertension? Anesth Analg 59: 435–437

— Winn HR, Tyson G, Park TS, Jane JA (1980b) Lidocaine prevents increased ICP after endotracheal intubation. In: Shulman K, Marmarou A, Miller JD,

Becker DP, Hochwald GM, Brock M (eds) Intracranial pressure IV. Springer, Berlin Heidelberg New York, pp 595–598

Bedford RF, Marshall WK, Butler A, Welsh JE (1981) Cardiac catheters for diagnosis and treatment of venous air embolism. J Neurosurg 55: 610–614

Belapavlovic M, Buchthal A (1982) Modification of ketamine-induced intracranial hypertension in neurosurgical patients by pretreatment with midazolam. Acta Anaesth Scand 26: 458–462

— — (1983) Intracranial pressure changes during the induction of anaesthesia with etomidate in neurosurgical patients. In: Isshii S, Nagai H, Brock M (eds) Intracranial Pressure V (eds). Springer, Berlin Heidelberg, pp 834–837

— — Beks JWF (1985) Barbiturates for cerebral aneurysm surgery. Acta Neurochir (Wien) 76: 73–81

— — (1986) Effect of isoflurane on intracranial pressure in patients with intracranial mass lesions. In: Miller JD, Teasdale GM, Rowan JO, Galbraith SL, Mendelow AD (eds) Intracranial pressure VI. Springer, Berlin Heidelberg, pp 725–731

Bendo AA, Kass IS, Cottrell JE (1987) Anesthetic protection against anoxic damage in the rat hippocampal slice. Brain Res 403: 136–141

Bendtsen AO, Cold GE, Saaby Lomholt B, Fischer-Christensen S, Hald A (1983) Central haemodynamic during thiopental load and sodium nitroprusside induced hypotension. Acta Anaesthesiol Scand [Suppl] 78: 82: Abstract no 123

— — Astrup J, Rosenørn J (1984) Thiopental loading during controlled hypotension for intracranial aneurysm surgery. Acta Anaesthesiol Scand 28: 473–477

— Kruse A, Madsen JB, Astrup J, Rosenørn J, Blatt-Lyon B, Cold GE (1985) Use of a continuous infusion of althesin in neuroanaesthesia. Changes in cerebral blood flow, cerebral metabolism, the EEG and plasma alphaxalone concentration. Br J Anaesth 57: 369–374

Bennett DR, Madsen JA, Jordan WS, Wiser WC (1973) Ketamine anesthesia in brain-damaged epileptics: electroencephalographic and clinical observations. Neurology (Minneap) 23: 449–460

Benthuysen JL, Kien ND, Quam DD, Martucci RW (1985) The influence of narcotic-induced rigidity on intracranial pressure. Anesthesiology 63: A393

Berne RM, Rubio R, Curnish RR (1974) Release of adenosine from ischemic brain. Effect on cerebral vascular resistance and incorporation into cerebral adenine nucleotides. Circ Res 35: 262–271

Bernstein JS, Nelson MA, Ebert TJ, Woods MP, Roerig DL (1987) Beat-by-beat cardiovascular responses to rapid sequence induction in humans: Effects of labetamol. Anesthesiology 67: A32

Berntman L, Welch FA, Bian Rosa IJ, Harp JR (1979) Diazepam fails to protect brain tissue in hypoxic stress. Anesthesiology 51: S202

Bingham RM, Procaccio F, Prior PF, Hinds CJ (1985) Cerebral electrical activity influences the effects of etomidate on cerebral perfusion pressure in traumatic coma. Br J Anaesth 57: 843–848

Bingham RM, Hinds CJ (1987) Influence of bolus doses of phenoperidine on intracranial pressure and systemic arterial pressure in traumatic coma. Br J Anaesth 59: 592–595

Black S, Ockert DB, Oliver WC, Cucchiara RF (1988) Outcome following posterior fossa craniotomy in patients in the sitting or horizontal positions. Anesthesiology 69: 49–56

Bleyaert AL, Nemoto EM, Safar P, Stezoski SW, Mickell JL, Moossy J, Rao GR (1978) Thiopental amelioration of brain damage after global ischaemia in monkeys. Anesthesiology 49: 390–398

Boardini DJ, Kassell NF, Coenter HC (1984) Comparison of sodium thiopental and methohexital for high-dose barbiturate anesthesia. J Neurosurg 60: 602–608

Boas RA, Covino BG, Shahnarian A (1982) Analgesic responses to i.v. lignocaine. Br J Anaesth 54: 501–504

Boheimer N, Ward S, Dopson T, Simmonds R, Weatherley B, Williams S (1984) Pharmacokinetics of laudanosine and a quaternary alcohol after an i.v. bolus dose of atracurium in patients with impaired renal function. Br J Anaesth 57: 345

Boop WC, Knight R (1978) Enflurane anesthesia and changes of intracranial pressure. J Neurosurg 48: 228–231

Botty C, Brown B, Stanley V et al (1968) Clinical experiences with the compound 347, a halogenated anesthetic compound. Anest Analg 47: 499–505

Bovill JG, Sebel PS, Wauquier A, Rog P (1982) Electroencephalographic effects of sufentanil anaesthesia in man. Br J Anaesth 54: 45–52

Bowdle TA, Artru AA (1988) Treatment of air embolism with a special pulmonary artery catheter introducer sheath in sitting dogs. Anesthesiology 68: 107–110

Braestrup C, Squires RF (1977) Specific benzodiazepine receptors in rat brain characterized by high affinity (3H)diazepam binding. Proc Natl Acad Sci USA 74: 3805–3809

Brandt L, Dick W, Erdmann K (1985) Nitrous oxide influences EEG changes induced by halothane, enflurane and isoflurane. Anesthesiology 63: A409

Branston NM, Hope DT, Symon L (1979) Barbiturates in focal ischemia of primate cortex: effects on blood flow distribution, evoked potential and extracellular potassium. Stroke 10: 647–653

Brian JE, McPherson RW, Traystman RJ (1988) Evolution of cerebral blood flow with time during 1.4 and 2.8% isoflurane in dog. Anesthesiology 69: A532

Bricolo AP, Glick RP (1981) Barbiturate effects on acute experimental intracranial hypertension. J Neurosurg 55: 397–406

Bristow A, Shalev D, Rice B, Lipton JM, Giesecke AH (1987) Low-dose synthetic narcotic infusions for cerebral relaxation during craniotomies. Anesth Analg 66: 413–416

Brodrick PM (1987) The sitting position; monitoring, diagnosis and treatment of air embolism. In: Jewkes DA (ed) Bailliere's Clinical Anesthesiology. Anaesthesia for Neurosurgery. WB Saunders, pp 419–440

— Ingram GS (1988) The antigravity suit in neurosurgery. Anaesthesia 43: 762–765

Brown SC, Lam AM, Manninen PH (1986) Haemodynamic effects of high-dose mannitol in man. Canad Anaesth Soc J 33: S92–S93

Bruce DA, Langfitt TW, Miller JD, Schutz H, Vapalahti M, Stanek A, Goldberg

HI (1973) Regional cerebral blood flow, intracranial pressure, and brain metabolism in comatose patients. J Neurosurg 38: 131–145

Buckland RW, Manners JM (1976) Venous air embolism during neurosurgery.
Anaesthesia 31: 633–643

Bullock R, van Dellen JR, Cambell D, Osborn I, Reinach SG (1986) Experience
with althesin in the management of persistently raised ICP following severe head
injury. J Neurosurg 64: 414–419

Bunegin L, Albin MS, Ruiz M (1984) Intracranial pressure responses following
rapid induction of hypotension with trimethapan. Anesthesiology 61: A369

—— Gelineau EF (1987) Effect of esmolol on cerebral blood flow during
intracranial hypertension and hemorrhagic hypovolemia. Anesthesiology 67:
A424

Bünemann L, Jensen K, Thomsen L, Riisager S (1987) Cerebral blood flow and
metabolism during controlled hypotension with sodium-nitroprusside and general anaesthesia for total hip replacement a.m. Charnley. Acta Anaesthesiol
Scand 31: 487–490

Burchiel KJ, Stockard JJ, Calverley RK, Smith NT (1977) Relationship of pre
and postanaesthetic EEG abnormalities to enflurane-induced seizure activity.
Anesth Analg 56: 509–514

Burke AM, Quest DO, Chien S, Cerri C (1981) The effects of mannitol on blood
viscosity. J Neurosurg 55: 550–553

Burton RR, Krutz RW (1975) G tolerance and protection with anti-G suit
concepts. Aviation Space Environm Med 46: 119–124

Butler BD, Hills BA (1979) The lung as a filter for microbubbles. J Appl Physiol
47: 537–543

Butler BD, Hills BA (1985) Transpulmonary passage of venous air emboli. J Appl
Physiol 59: 543–547

— Luehr S, Hills B, Katz J (1985) Nitrous oxide anesthesia and pulmonary air
embolism. Anesthesiology 63: A422

— Leiman BC, Luehr S, Katz J (1986) Effects of PEEP on the incidence of
paradoxical air embolism in the absence of ASD in dogs. Anesthesiology 65: A81

Calliauw L, Van Aken J, Rolly G, Verbeke L (1987) The position of the patient
during neurosurgical procedures on the posterior fossa. Acta Neurochir (Wien)
85: 154–158

Campan L, Lazorthes Y (1976) Note sur les effects comparés des benzodiazépines
et de la chlorpromazine sur la pression intracranienne du chien. Ann Anesthesiol
Fr 17: 1193–1198

Campkin TV (1984) Isoflurane and cranial extradural pressure, A study in neurosurgical patients. Br J Anaesth 56: 1083–1087

— Flinn RM (1986) Isoflurane: Its use to induce hypotension in neurosurgical
patients. Europ J Anesthesiology 3: 395–401

—— (1989) Isoflurane and cerebrospinal fluid pressure. A study in neurosurgical
patients undergoing intracranial shunt procedures. Anaesthesia 44: 50–54

Carlsson C, Hägerdal M, Siesjö BK (1975a) Increase in cerebral oxygen uptake
and blood flow in immobilization stress. Acta Physiol Scand 95: 206–208

— Harp JR, Siesjö BK (1975b) Metabolic changes in the cerebral cortex of the rat induced by intravenous pentothalsodium. Acta Anaesthesiol Scand [Suppl] 57: 7–17

— Hägerdal M, Siesjö BK (1976a) The effect of nitrous oxide on oxygen consumption and blood flow in the cerebral cortex of the rat. Acta Anaesthesiol Scand 20: 91–95

— — Kaasik AE, Siesjö BK (1976b) The effects of diazepam on cerebral blood flow and oxygen consumption in rats and its synergistic interaction with nitrous oxide. Anesthesiology 45: 319–325

— Chapman AG (1981) The effects of diazepam on the cerebral metabolic state in rats and its interaction with nitrous oxide. Anesthesiology 54: 488–495

— Smith DS, Keykhah MM, Englebach I, Harp JR (1982) The effects of high-dose fentanyl on cerebral circulation and metabolism in rats. Anesthesiology 57: 375–380

Carpenter RL, Eger EI II, Johnson BH, Unadkat JD, Sheiner LB (1986) Pharmacokinetics of inhaled anesthetics in humans: Measurements during and after the simultaneous administration of enflurane, halothane, isoflurane, methoxyplurane and nitrous oxide. Anesth Analg 65: 575–582

Carter LP, Atkinson JR (1973) Cortical blood flow in controlled hypotension as measured by thermal diffusion. J Neurol Neurosurg Psychiatry 36: 906–913

Cartwright P, Prys-Roberts C, Gill K, Stafford ADM, Gray A (1983) Ventilatory depression related to plasma fentanyl concentrations during and after anesthesia in humans. Anesth Analg 62: 966–974

Casthely PA, Lear S, Cottrell JE, Lear E (1982) Intrapulmonary shunting during induced hypotension. Anesth Analg 61: 231–235

Cavazzuti M, Porro CA, Biral GP, Benassi C, Barbieri GC (1987) Ketamine effects on local cerebral blood flow and metabolism in the rat. J Cereb Blood Flow Metab 7: 806–811

Chang JL, Chang GL, Nemoto EM, Hung TK, Nemmer JP (1986) Protective effect of i.v. lidocaine in acute experimental spinal cord injury. Anesthesiology 65: A319

Chapman AG, Nordström C-H, Siesjö BK (1978) Influence of phenobarbital anesthesia on carbohydrate and amino-acid metabolism in rat brain. Anesthesiology 48: 175–182

Chapman CR, Benedetti C (1979) Nitrous oxide effects on cerebral evoked potential to pain. Anesthesiology 51: 135–138

Chapple DJ, Miller AA, Ward JB, Wheatley PL (1987) Cardiovascular and neurological effects of laudanosine. Br J Anaesth 59: 218–225

Chen RYZ, Matteo RS, Fan F, Schuessler GB, Chien S (1982a) Resetting of baroreflex sensitivity after induced hypotension. Anesthesiology 56: 29–35

— Fan FC, Schuessler GB, Chien S (1982b) Distribution of cerebral blood flow (CBF) during halothane anesthesia. Anesthesiology 57: A37

— — Carlin RD, Schuessler GB, Chien S (1984) Comparison of regional cerebral blood flow during isoflurane and halothane induced hypotension. Anesthesiology 61: A21

Cheng S-C, Brunner EA (1981) Inhibition of GABA metabolism in rat brain slices by halothane. Anesthesiology 55: 26–33

Chiache M, Fukunaga AF, Bloor BC, van Etten A (1983) A comparative study on the sympathetic and the metabolic activities during induced hypotension with adenosine triphosphate and sodium nitroprusside. Anesthesiology 59: A10

Chiolero RL, Ravussin P, Anderes JP, de Tribolet N, Freeman J (1986a) Midazolam reversal with RO-15-1788 in patients with severe head injury. Anesthesiology 65: A358

— — Chassot PG, Neff B, Freeman J (1986b) RO-15-1788 for rapid recovery after craniotomy. Anesthesiology 65: A466

Choi DW, Farb DH, Fischbach GD (1977) Chlordiazepoxide selectively augments GABA action in spinal cord cultures. Nature 269: 342–344

Chraemmer-Jørgensen B, Høllund-Carlsen PF, Marving J, Christensen V (1986) Lack of effect of intravenous lidocaine on hemodynamic responses to rapid sequence induction of general anesthesia: A double-blind clinical trial. Anesth Analg 65: 1037–1041

Christensen MS, Høedt-Rasmussen K, Lassen NA (1967) Cerebral vasodilatation by halothane anaesthesia in man and its potentiation by hypotension and hypercapnia. Br J Anaesth 927–934

Church J, Zeman S, Lodge D (1988) The neuroprotective action of ketamine and MK-801 after transient cerebral ischemia in rats. Anesthesiology 69: 702–709

Clark DL, Hosick EC, Adams N et al (1973) Neural effects of isoflurane (Forane) in man. Anesthesiology 39: 261–270

Clayton DG, Evans P, Williams C, Thurlow AC (1985) Paradoxical air embolism during neurosurgery. Anaesthesia 40: 981–989

Cohen AT, Kelly DR (1987) Assessment of alfentanil by intravenous infusion as long-term sedation in intensive care. Anaesthesia 42: 545–548

Cohen EN, Chow KL, Mathers L (1972) Autoradiographic distribution of volatile anesthetics within the brain. Anesthesiology 37: 324–331

Cohen PJ, Wollman H, Alexander SC, Chase PE, Behar MG (1964) Cerebral carbohydrate metabolism in man during halothane anesthesia. Effects of $PaCO_2$ on some aspect of carbohydrate utilization. Anesthesiology 25: 185–191

Cold GE (1975) Nitrous oxide and intracranial pressure. Br J Anaesth 47: 1119

— Eskesen V, Eriksen H, Amtoft O, Madsen JB (1985) CBF and $CMRO_2$ during continuous etomidate infusion supplemented with N_2O and fentanyl in patients with supratentorial cerebral tumours. A dose response study. Acta Anaesthesiol Scand 29: 490–494

— — — Blatt Lyon B (1986) Changes in $CMRO_2$, EEG and concentration of etomidate in serum and brain tissue during craniotomy with continuous etomidate supplemented with N_2O and fentanyl. Acta Anaesthesiol Scand 30: 159–163

— Christensen KJS, Nordentoft J, Engberg M, Bach Pedersen M (1988) Cerebral blood flow, cerebral metabolic rate of oxygen and relative CO_2 reactivity during neurolept anesthesia in patients subjected to craniotomy for supratentorial cerebral tumors. Acta Anaesthesiol Scand 32: 310–315

Cold GE, Sinding H, Noreng M. Jugular venous bulb pressure in the sitting position. Effects of PEEP, anti-G suit pressure, and neck compression. Europ Congr Anaesth. 1990

Cole DJ, Drummond JC, Shapiro HM (1987) A comparison of the extent of ischemia following middle cerebral artery occlusion during three induced hypotensive techniques. Anesthesiology 67: A571

— Shapiro HM (1989) Different 1.2 MAC combinations of nitrous oxide-enflurane cause unique cerebral and spinal cord metabolic responses in the rat. Anesthesiology 70: 787–792

Colley PS, Cheney FW (1977) Sodium nitroprusside increases Qs/Qt in dogs with regional atelectasis. Anesthesiology 47: 338–341

Colley PS, Sivarajan M (1984) Regional blood flow in dogs during halothane anesthesia and controlled hypotension produced by nitroprusside or nitroglycerin. Anesth Analg 63: 503–510

Colpaert FC, Niemegeers CJE, Janssen PAJ, Maroli AN (1980) The effects of prior fentanyl administration and of pain on fentanyl analgesia. Tolerance to the enhancement of narcotic analgesia. J Pharmacol Exp Ther 213: 418–424

Corkill G, Chikovanni OI, McLeish I, McDonald LW, Youmans JR (1976) Timing of pentobarbital administration for brain protection in experimental stroke. Surg Neurol 5: 147–149

— Sivalingham S, Reital JA, Gilroy BA, Helphrey MG (1978) Dose dependency of the post-insult protective effect of pentobarbital in the canine experimental stroke model. Stroke 9: 10–12

Corssen G, Domino EF, Bree RL (1969a) Electroencephalographic effects of ketamine anesthesia in children. Anesth Analg 48: 141–147

— Groces EH, Gomez S, Allen RJ (1969b) Ketamine: its place in anesthesia for neurosurgical diagnostic procedures. Anesth Analg 48: 181–188

Cote CJ, Greenhow E, Marshall BE (1979) The hypotensive response to rapid intravenous administration of hypertonic solutions in man and in the rabbit. Anesthesiology 50: 30–35

Cotev S, Shalit MN (1975) Effects of diazepam on cerebral blood flow and oxygen uptake after head injury. Anesthesiology 43: 117–122

Cottrell JE, Robustelli A, Post K, Turndorf H (1977) Furosemide- and mannitol-induced changes in intracranial pressure and serum osmolality and electrolytes. Anesthesiology 47: 28–30

— Casthely P, Brodie JD, Patel K, Klein A, Turndorf H (1978) Prevention of nitroprusside-induced cyanide toxicity with hydroxocobalamin. N Engl J Med 298: 809–811

— Gupta B, Rappaport H, Turndorf H, Ransohoff J, Flamm ES (1980) Intracranial pressure during nitroglycerin-induced hypotension. J Neurosurg 53: 309–311

— Giffin JP, Lim K, Milhorat T, Stein S, Shwiry B (1982) Intracranial pressure, mean arterial pressure and heart rate following midazolam or thiopental in humans with intracranial masses. Anesthesiology 57: A323

— Hartung J, Giffin JP, Shwiry B (1983) Intracranial and hemodynamic changes after succinylcholine administration in cats. Anesth Analg 62: 1006–1009

Criado A, Maseda J, Navarro E, Escarpa A, Avello F (1980) Induction of anaesthesia with etomidate: Haemodynamic study of 36 patients. Br J Anaesth 52: 803–805

Crockard HA, Brown FD, Mullan JF (1976) Effects of trimethapan and sodium nitroprusside on cerebral blood flow in rhesus monkeys. Acta Neurochir (Wien) 35: 85–89

Crosby G, Crane AM, Sokoloff L (1982) Local changes in cerebral glucose utilization during ketamine anaesthesia. Anaesthesiology 56: 437–443

— — — (1984) A comparison of local rates of glucose utilization in spinal cord and brain in conscious and nitrous oxide- or pentobarbital-treated rats. Anesthesiology 61: 434–438

Crozier WC, Henney JB, Rogers MC, Traystman RJ (1986) Persistence of cerebral blood flow autoregulation during nitroprusside administration. Anesthesiology 65: A574

Crumrine RS, Nulsen FE, Wess MH (1975) Alterations in ventricular fluid pressure during ketamine anesthesia in hydrocephalic children. Anesthesiology 42: 758–761

Cucchiara RF, Theye RA, Michenfelder JD (1974) The effects of isoflurane on canine cerebral metabolism and blood flow. Anesthesiology 40: 571–574

Cunitz G, Danhauser I, Wickbold J (1973) Comparative investigations on the influence of etomidate, thiopentone and methohexitone on the intracranial pressure of the patient. Anaesthesist 22: 357–366

— — Gruss P (1976) Die Wirkung von Enflurane (Etrane) im Vergleich zu Halothan auf den intracraniellen Druck. Anaesthesist 25: 323–330

Dahlgren N, Ingvar M, Yokoyama H, Siesjö BK (1981) Effect of indomethacin on local cerebral blood flow in awake, minimally restrained rats. J Cereb Blood Flow Metab 1: 233–236

Dalrymple DG, MacGowan SW, MacLeod GF (1979) Cardiorespiratory effects of the sitting position in neurosurgery. Br J Anaesth 51: 1079–1082

Darimont PC, Jenkins LC (1977) The influence of intravenous anesthetics on enflurane-induced central nervous system seizure activity. Can Anaesth Soc J 24: 42–56

Davies DW, Kadar D, Steward DJ, Munro IR (1975) A sudden death associated with the use of sodium nitroprusside for induction of hypotension during anaesthesia. Can Anaesth Soc J 22: 547–552

Davis DW, Hawkins RA, Mans AM, Hibbard LS, Biebuyck JF (1984) Regional cerebral glucose utilization during althesin anesthesia. Anesthesiology 61: 362–368

— Mans AM, Biebuyck JF, Hawkins RA (1986) Regional brain glucose utilization in rats during etomidate anesthesia. Anesthesiology 64: 751–757

— — — Phil CBD, Hawkins RA (1988) The influence of ketamine on regional brain glucose use. Anesthesiology 69: 199–205

Davis RF, Douglas ME, Heenan TJ, Downs JB (1981) Brain tissue pressure measurements during sodium nitroprusside infusion. Crit Care Med 9: 17–21

Dawson B, Michenfelder JD, Theye RA (1971) Effect of ketamine on canine cerebral blood flow and metabolism: Modification by prior administration of thiopental. Anesth Analg 50: 443–447

Day AL, Friedman WA, Sypert GW, Mickle JP (1982) Successful treatment of the normal perfussion pressure breakthrough syndrome. Neurosurgery 11: 625–630

Dearden NM, McDowall DG (1985) Comparison of etomidate and althesin in the reduction of increased intracranial pressure after head injury. Br J Anaesth 57: 361–368

DeCastro J, VandeWater A, Wouters L, Xhonneux R, Reneman R, Kay B (1979) Comparative study of cardiovascular neurological and metabolic side-effects of eight narcotics in dogs. Acta Anaesthesiol Belg 30: 5–99

DeGuerra O, Bunegin L, Albin MS (1986) Changes in cerebral blood flow (CBF) following rapid induction of hypotension with sodium nitroprusside. Crit Care Med 388

De Jong RH, Heavner JE (1971) Correlation of the Ethrane Electroencephalogram with motor activity in cats. Anesthesiology 35: 474–481

Delaney TJ, Miller ED (1980) Rebound hypertension after sodium nitroprusside prevented by saralasin in rats. Anesthesiology 52: 154–156

DeRood M, Deloof T, Berre J, Verbist A, Frühling J, Phuoc T (1980) Effect of 1% enflurane anesthesia on cerebral blood flow and metabolism in neurosurgical patients during normo- and hyperventilation. Acta Anaesthesiol Belg 31 [Suppl]: 3–19

DeValois JC, Peperkamp JPC (1971) The influence of some drugs upon the regulation of cerebral blood flow in the rabbit. In: Ross Russell (ed) Brain and blood flow. Pitman Medical and Scientific Publishing Co LTD, London, pp 254–257

DiGiovanni AJ, Goodrick J, Neigh JL, Harp JR, Kennell EM (1974) The effect of halothane anesthesia on intracranial pressure in the presence of intracranial hypertension. Anesth Analg 53: 823–828

Doenicke A, Lorenz W, Beigl R, Bezecny H, Uhlig G, Kalmar L, Praetorius B, Mann G (1973) Histamine release after intravenous application of short-acting hypnotics. A comparison of etomidate, althesin and propanidid. Br J Anaesth 45: 1097–1104

— Kugler J, Penzel G et al (1973) Hirnfunktion und toleranzbreite nach etomidate, einem neuen, barbituratfreien i.v. applizierbaren hypnotikum. Anaesthesist 22: 357–366

— Löffler B, Kugler J, Suttmann H, Grote B (1982) Plasma concentration and EEG after various regimens of etomidate. Br J Anaesth 54: 393–400

Dohl S, Matsumoto M, Takahashi T (1981) The effects of nitroglycerin on cerebrospinal fluid pressure in awake and anaesthetized humans. Anesthesiology 54: 511–514

Donegan MF, Bedford RF (1980) Intravenously administered lidocaine prevents intracranial hypertension during endotracheal suctioning. Anesthesiology 52: 516–518

Drummond JC, Todd MM, Toutant SM, Shapiro HM (1983a) Brain surface protrusion during enflurane, halothane and isoflurane anesthesia in cats. Anesthesiology 59: 288–293

— — Shapiro HM (1983b) Cerebral blood flow autoregulation in the cat during anesthesia with halothane and isoflurane. Anesthesiology 55: A305

Drummond JC, Todd MM (1985) The response of feline cerebral circulation to PaCO$_2$ during anesthesia with isoflurane and halothane and during sedation with nitrous oxide. Anesthesiology 62: 268–273

— — Scheller MS, Shapiro HM (1986) A comparison of the direct cerebral vasodilating potencies of halothane and isoflurane in the New Zealand White Rabbit. Anesthesiology 65: 462–467

— Scheller MS, Todd MM (1987) The effect of nitrous oxide on cortical blood flow during anesthesia with halothane and isoflurane, with and without morphine, in the rabbit. Anesth Analg 66: 1083–1089

Dubois MY, Sato S, Chassy J, MacNamara TE (1982) Effects of enflurane on brainstem auditory evoked responses in humans. Anesth Analg 61: 898–902

Durieux ME, Monk CR, Sperry RJ, Matthern GE (1988) Labetalol preserves blood flow to vital organs during deliberate hypotension induced by isoflurane. Anaesthesiology 69: A899

Ebrahim ZY, DeBoer GE, Luders H, Hahn JF, Lesser RP (1986) Effect of etomidate on the electroencephalogram of patients with epilepsia. Anesth Analg 65: 1004–1006

Edde RR, Smalley S (1979) Defect in oxygenation associated with mannitol. Anesth Analg 58: 145–146

Edmonds-Seal J, Prys-Roberts C, Adams AP (1971) Air embolism, a comparison of various methods of detection. Anaesthesia 26: 202–208

Edmondson RH, Delvalle OE, Shah N, Coffey CR, Dwyer D, Matarazzo D, Thorne A, Sogani P, Herr H, Wong G (1988) Esmolol infusion during nitroprusside-induced hypotension: Impact on the cardiovascular, renin-angiotensin and sympathetic nervous systems. Anesthesiology 69: A36

Edwards JE (1968) Interatrial communication. In: Gould SE (ed) Pathology of the heart. Ch C Thomas, Springfield, Jee, pp 260–261

Eger EI, Stevens WC, Cromwell TH (1971) The electroencephalogram in man anesthetized with forene. Anesthesiology 35: 504–508

— (1984) The pharmacology of isoflurane. Br J Anaesth 56: 71S–99S

— (1981) Isoflurane: A review. Anesthesiology 55: 559–576. Eger EI

Eintrei C, Leszniewski W, Carlsson C (1985) Local application of 133-Xenon for measurement of regional cerebral blood flow (rCBF) during halothane, enflurane and isoflurane anesthesia in humans. Anesthesiology 63: 391–394

— Carlsson C (1986) Effects of hypotension induced by adenosine on brain surface oxygen pressure and cortical cerebral blood flow in the pig. Acta Physiol Scand 126: 463–469

Eisenberg HM, Frankowski RF, Contant CF, Marshall LF, Walker MD (1988) High-dose barbiturate control of elevated intracranial pressure in patients with severe head injury. J Neurosurgery 69: 15–23

Engberg M, Øberg B, Christensen KS, Bach Pedersen M, Cold GE (1989) The arterio-venous oxygen content differences (AVDO$_2$) during halothane and neurolept anaesthesia in patients subjected to craniotomy. Acta Anaesthiol Scand 33: 642–646

Engberg M, Melsen NC, Herlevsen P, Haraldsted V, Cold GC (1990) Changes of blood pressure and cerebral arterio-venous oxygen content differences (AVDO$_2$)

with and without bupivacaine scalp infiltration during craniotomy. Acta Anaesthesiol Scand (in press)

English JB, Westenskow D, Hodges MR, Stanley TH (1978) Comparison of venous air embolism monitoring methods in supine dogs. Anesthesiology 48: 425–429

Erdmann K, Brandt L (1986) Paradoxical EEG arousal phenomenon following anesthesia with isoflurane. Anesthesiology 65: A347

Ericsson JA, Gottlieb JD, Sweet RB (1964) Closed chest massage in the treatment of venous air embolism. New Eng J Med 270: 1353–1354

Ernst PS, Albin MS, Bunegin L, Levin D, Gelineau J (1988) Cerebrovascular and spinal cord dynamics of the sitting position. Anesthesiology 69: A554

Evans DE, Kobrine AI, Legrys DC, Bradley ME (1984) Protective effect of lidocaine in acute cerebral ischemia induced by air embolism. J Neurosurg 60: 257–263

— — (1987) Reduction of experimental intracranial hypertension by lidocaine. Neurosurgery 20: 542–547

— Catron PW, McDermott JJ, Thomas LB, Kobrine AI, Flynn ET (1989) Effect of lidocaine after experimental cerebral ischemia induced by air embolism. J Neurosurg 70: 97–102

Evans J, Rosen M, Weeks RD, Wise C (1971) Ketamine in neurosurgical procedures. Lancet i: 40–41

Ewing JR, Welch KMA, Robertson WM, Brown GG, Diaz FG, Ausman J (1983) A probe-by-probe identification of focal cerebral ischemia using the 133-Xe inhalation technique. J Cereb Blood Flow Metab 3: 586–587

Faden AI, Jacobs TP, Holaday JW (1981a) Opiate antagonist improves neurologic recovery after spinal injury. Science 211: 493–494

— — Mougey E, Holoday JW (1981b) Endorphins in experimental spinal injury: Therapeutic effect of nalaxone. Ann Neurol 10: 227–231

— — Holady JW (1982a) Comparison of early and late nalaxone treatment in experimental spinal injury. Neurology 32: 677–681

— Hallenbeck JM, Brown CQ (1982b) Treatment of experimental stroke: comparison of nalaxone and thyreotropin releasing hormone. Neurology 32: 1083–1087

Fahey MR, Rupp SM, Fisher DM, Miller RD, Sharma M, Canfell C, Castagnoli K, Hennis PJ (1984) The pharmacokinetics and pharmacodynamics of atracurium in patients with and without renal failure. Anesthesiology 61: 699–714

Fahmy NR, Gavras HP (1983) Captopril decreases nitroprusside requirements and prevents rebound hypertension following cessation of nitroprusside infusion in humans. Anesthesiology 59: A359

— (1985) Nitroprusside vs. a nitroprusside–trimethaphan mixture for induced hypotension: Hemodynamic effects and cyanide release. Clin Pharmacol Ther 37: 264–270

— Bottros M, Dimedale J, Gaivin RJ, Lucas V (1986) A comparison of the hemodynamic and hormonal effects of a nitroprusside–trimethaphan mixture with nitroprusside or trimethapan alone for induced hypotension. Anesthesiology 65: A70

Feig U, McCurdy DK (1977) The hypertonic state. New Engl J Med 26: 1444–1454

Ferrer-Allado T, Brechner VL, Dymond A, Cozen H, Crandall P (1973) Ketamine-induced electroconvulsive phenomenon in the human limbic and thalamic regions. Anesthesiology 38: 333–344

Feustel PJ, Ingvar MC, Severinghaus JW (1981) Cerebral oxygen availability and blood flow during middle cerebral artery occlusion: Effects of pentobarbital. Stroke 12: 858–863

Finlay WE, McKee JI (1982) Screen cortisol levels in severely stressed patients. Lancet i: 1414–1415

Fitch W, Barker J, Jennett WB, McDowall DG (1969) The influence of neurolept-analgesic drugs on cerebrospinal fluid pressure. Br J Anaesth 41: 800–806

— McDowall DG (1971) Effect of halothane on intracranial pressure gradients in the presence of intracranial space-occupying lesions. Br J Anaesth 43: 904–911

— MacKenzie ET, Harper AM (1975) Effects of decreasing arterial blood pressure on cerebral blood flow in the baboon: influence of sympathetic nervous system. Circ Res 37: 550–557

— Ferguson GG, Sengupta D, Garibi J, Harper AM (1976) Autoregulation of cerebral blood flow during controlled hypotension in baboons. J Neurol Neurosurg Psychiatry 39: 1014–1022

— (1977) Editorial. Sodium nitroprusside and the cerebral circulation. Br J Anaesth 49: 399–400

Fitch W, McGeorge AP, MacKenzie ET (1978) Anaesthesia for studies of the cerebral circulation: A comparison of phencyclidine and althesin in the baboon. Br J Anaesth 50: 985–990

Fitzpatrick JH, Gilboe DD (1982) Effects of nitrous oxide on the cerebrovascular tone, oxygen metabolism, and electroencephalogram of the isolated perfused canine brain. Anesthesiology 57: 480–484

Flacke JW, Flacke WE, Williams GD (1977) Acute pulmonary edema following nalaxone reversal of high-dose morphine anesthesia. Anesthesiology 47: 376–378

Flamm ES, Demopoulos HB, Seligman ML, Ransohoff J (1977) Possible molecular mechanisms of barbiturate-mediated protection in regional cerebral ischaemia. Acta Neurol Scand [Suppl] 64: 150–151

— Young W, Demopoulus HB, Decrescito V, Tomasula JJ (1982) Experimental spinal cord injury: treatment with nalaxone. Neurosurgery 10: 227–231

Fleischer JE, Milde JH, Moyer TP, Michenfelder JD (1988) Cerebral effects of high-dose midazolam and subsequent reversal with RO 15-1788 in dogs. Anesthesiology 68: 234–242

Fontenot HJ, Wilson RD, Norris JC, Ho IK (1984) The GABA system: New evidence of neurotransmitter involvement in the mechanism of anesthesia. Anesthesiology 61: A327

Ford EW, Morrell F, Whisler WW (1982) Methohexital anesthesia in the surgical treatment of uncontrollable epilepsy. Anesth Analg 61: 997–1001

Forrester T, Harper AM, MacKenzie ET, Thomson EM (1979) Effect of adenosine trophosphate and some derivates on cerebral blood flow and metabolism. J Physiol 296: 343–355

Forster A, VanHorn K, Marshall LF, Shapiro HM (1978) Anesthetic effects on blood–brain barrier function during acute arterial hypertension. Anesthesiology 49: 26–30

— Juge O, Morel D (1982) Effects of midazolam on cerebral blood flow in human volunteers. Anesthesiology 56: 453–455

— — Louis M, Nahory A (1987) Effects of a specific benzodiazepine antagonist (RO 15-1788) on cerebral blood flow. Anesth Analg 66: 309–313

Francis A, Pulsinelli WA (1983) Increased binding of (3H)GABA to striatal membranes following ischemia. J Neurochem 40: 1497–1499

Fragen RJ, Shanks CA, Molteni A, Avram MJ (1984) Effects of etomidate on hormonal responses to surgical stress. Anesthesiology 61: 652–656

Frank M, Savege TM, Leigh M, Greenwood J, Holly JNP (1982) Comparison of the cerebral function monitor and plasma concentrations of thiopentone and alphaxolone during total i.v. anaesthesia with repeated bolus doses of thiopentone and althesin. Br J Anaesth 54: 609–616

Freeman J, Ingvar DH (1967) Effects of fentanyl on cerebral cortical blood flow and EEG in the cat. Acta Anaesth Scand 11: 381–391

Freuchen IP (1959) The use of an anti-gravity suit in neurosurgery. Acta Anaesth Scand 3: 17–23

Freye E, Hartung E, Klatte A, Abel J (1983) Plasma levels of alfentanil and etomidate in patients and their relation to compressed power spectral analysis of the EEG. Acta Anaesthesiol Belg 33: 87–96

Friesen RH, Thieme RE, Hondas AT, Morrison JE (1987) Changes in anterior fontanel pressure in preterm neonates receiving isoflurane, halothane, fentanyl or ketamine. Anesth Analg 66: 431–434

Frost EAM (1984) Inhalation anaesthetic agents in neurosurgery. Br J Anaesth 56: 47S–56S

Fukunaga AF, Olewine SK, van Etten AP (1981) Hemodynamic effects of ATP and nitroprusside. Anesthesiology 53: A13

Fukunaga AF, Flacke WE, Bloor BC (1982) Hypotensive effects of adenosine and adenosine Triphosphate compared with sodium nitroprusside. Anesth Analg 61: 273–278

Fukunaga AF, Sodeyama O, Matsuzaki Y (1983) Hemodynamic and metabolic changes of ATP-induced hypotension during surgery. Anesthesiology 59: A12

Fuller J, Gelb AW, Karlik S (1980) The influence of halothane and innovar on brain oedema formation. Can Anesth Soc J S110

Gancher S, Laxer KD, Krieger W (1984) Activation of epileptogenic activity by etomidate. Anesthesiology 61: 616–618

Garcia-Bengochea F, Munson ES, Freeman JV (1976) The lateral sitting position for neurosurgery. Anesth Analg 55: 326–330

Gardner AE, Olson BE, Lichtiger M (1971) Cerebrospinal-fluid pressure during dissociative anaesthesia with ketamine. Anesthesiology 35: 226–228

Geevarghese KP (1977) Anaesthetic management of patients undergoing surgery for posterior fossa lesions. Intern Anaesth Clin 15: 165–194

Gelb AW, Floyd P, Lok P, Peerless SJ, Farrell M (1986) A prophylactic bolus of thiopentone does not protect against prolonged focal ischaemia. Can Anesth Soc J 33: 173–177

— Steinberg GK, Lam AM, Manninen PH, Peerless SJ, Rassi-Neto A (1988) The effects of a bolus of lidocaine in focal cerebral ischaemia. Can J Anaesth 35: 489–493

Gelman S, Fowler KC, Smith LR (1984) Regional blood flow during isoflurane and halothane anaesthesia. Anesth Analg 63: 557–565

Gerson JI, Allen FB, Seltzer JL, Parker FB, Makowitz AH (1982) Arterial and venous dilatation by nitroprusside and nitroglycerin—Is there a difference? Anesth Analg 61: 256–260

Ghani GA, Sung YF, Weinstein MS, Tindall GT, Fleischer AS (1983) Effects of intravenous nitroglycerin on the intracranial pressure and volume pressure response. J Neurosurg 58: 562–565

Ghoneim MM, Yamada T (1977) Etomidate: A clinical and electroencephalographic comparison with thiopental. Anesth Analg 56: 479–485

Gibs JM (1972) The effect of intravenous ketamine on cerebrospinal fluid pressure. Br J Anaesth 44: 1298–1302

Gibson GE, Duffy TE (1981) Impaired synthesis of acetylcholine by mild hypoxia or nitrous oxide. J Neurochem 36: 28–33

Giffin JP, Cottrell JE, Hartung J, Shwiry B (1983) Intracranial pressure during nifedipine-induced hypotension. Anesth Analg 62: 1078–1080

— Litwak B, Cottrell JE, Hartung J, Capuano C (1985) Intracranial pressure, mean arterial pressure and heart rate after rapid paralysis with atracurium in cats. Can Anaesth Soc J 32: 618–621

— Hartung J, Cottrell JE, Capuano C, Shwiry B (1986) Effect of vecuronium on intracranial pressure mean arterial pressure and heart rate in cats. Br J Anaesth 58: 441–443

Gisvold SE, Safar P, Hendrick HHL, Rao G, Moossy J, Alexander H (1984) Thiopental treatment after global ischemia in pigtailed monkeys. Anesthesiology 60: 88–96

Glenski JA, Cucchiara RF, Michenfelder JD (1986) Transesophageal echocardiography and transcutaneous O_2 and CO_2 monitoring for detection of venous air embolism. Anesthesiology 64: 541–545

Goldstein A Jr, Wells BA, Keats AS (1966) Increased tolerance to cerebral anoxia by pentobarbital. Arch Int Pharmacodyn Ther 161: 138–143

Gordon E (1970) The action of drugs on intracranial contents. In: Boulton TB, Bryce-Smith R et al (eds) Progress in anaesthesiology. Excerpta medica, Amsterdam, p 60

— (1974) Anaesthesia for neurosurgery. In: Emeric Gordon (ed) A basis and practice of neuroanaesthesia. Excerpta medica, Amsterdam, pp 173–198

— Lagerkranser M, Rudehill A, von Host H (1988) The effect of isoflurane on cerebrospinal fluid pressure in patients undergoing neurosurgery. Acta Anaesthesiol Scand 32: 108–112

Grant IS, Hutchison G (1983) Epileptiform seizures during prolonged etomidate sedation. Lancet ii: 511–512

— (1986) Delayed convulsions following enflurane anaesthesia. Anaesthesia 41: 1024–1025

Greenwood J, Luthert PJ, Pratt OE, Lantos PL (1988) Hyperosmolar opening of the blood–brain barrier in energy-depleted rat brain. Part I. Permeability studies. J Cereb Blood Flow Metab 8: 9–15

Gregory P, Ishikawa T, McDowall DG (1981) CO_2 responses of the cerebral circulation during drug-induced hypotension in the cat. J Cereb Blood Flow Metab 1: 195–201

Griffith DPG, Cummins BH, Greenbaum R, Griffith HB, Staddon GE, Wilkins DG, Zorab JSM (1974) Cerebral blood flow and metabolism during hypotension induced with sodium nitroprusside. Br J Anaesth 46: 671–679

Gronert GA, Messick JM, Cucchiara RF, Michenfelder JD (1979) Paradoxical air embolism from a patent foramen ovale. Anesthesiology 50: 548–549

— Michenfelder JD, Sharbrough FW, Milde JH (1981) Canine cerebral metabolic tolerance during 24 hours deep pentobarbital anesthesia. Anesthesiology 55: 110–113

Gross PM, Harper AM, Teasdale GM (1981) Cerebral circulation and histamine: 1. Participation of vascular H_1 and H_2 receptors in vasodilatory responses to carotid artery infusion. J Cereb Blood Flow Metab 1: 97–108

Grosslight K, Foster R, Colohan AR, Bedford RF (1985) Isoflurane for neuroanesthesia: Risk factors for increases in intracranial pressure. Anesthesiology 63: 533–536

Grubb RL, Raichle ME (1982) Effect of hemorrhagic and pharmacologic hypotension on cerebral oxygen utilization and blood flow. Anesthesiology 56: 3–8

Grummitt RM, Goat VA (1984) Intracranial pressure after phenoperidine. Anaesthesia 39: 565–567

Guggiara M, Lechat PH, Garen C, Darnat S, Viars P (1985) Prevention of paradoxical air embolism by 2-D contrast echocardiography in neurosurgical patients. Anesthesiology 63: A425

Hagan CH, Marshall WK, Weber DM et al (1983) Succinylcholine increases CO_2: Neuroanaesthetic implications. Anesthesiology 59: A303

Hägerdal M, Welch FA, Keykhah MM, Perez E, Harp JR (1978) Protective effects of combinations of hypothermia and barbiturates in cerebral hypoxia in the rat. Anesthesiology 49: 165–169

— Keykhah MM, Perez E, Harp JR (1979) Additive effects of hypothermia and phenobarbital upon cerebral oxygen consumption in the rat. Acta Anaesthesiol Scand 23: 89–92

Haghighi SS, Chehrazi BB, Higgins RS, Remington WJ, Wagner FC (1987) Effect of lidocaine treatment on acute spinal cord injury. Neurosurgery 20: 536–541

Halldin M, Wåhlin Å (1959) Effect of succinylcholine on the intraspinal fluid pressure. Acta Anaesth Scand 3: 155–161

Hamby WB, Terry RN (1952) Air embolism in operations done in the sitting position. A report of five fatal cases and one of rescue by a simple maneuver. Surgery 31: 212–215

Hamill JF, Berford RF, Weaver DC, Colohan AR (1981) Lidocaine before endotracheal intubation: Intravenous or laryngotracheal? Anesthesiology 55: 578–581

Hankinson HL, Smith AL, Nielsen SL, Hoff JT (1974) Effect of thiopental on focal cerebral ischaemia in dogs. Surg Forum 25: 445–447

Hans P, Dethier JC, Godin D, Stevenaert A (1980) Compared effects of enflurane and of halothane on the intracranial pressure and the cerebral perfusion pressure in dog. Acta Anaesthesiol Belg 31 [Suppl]: 49–59

Hansen TD, Warner DS, Todd MM, Vust LJ, Trawick DC (1988a) Distribution of cerebral blood flow during halothane versus isoflurane anesthesia in rats. Anesthesiology 69: 332–337

— — — — — (1988b) Evidence for persistent flow-metabolism coupling during halothane vs isoflurane anesthesia. Anesthesiology 69: A912

— — — (1988c) Nitrous oxide is a more potent cerebral vasodilator than either halothane or isoflurane. Anesthesiology 69: A537

Haraldsted V, Asmussen J, Herlevsen P, Cold GE (1989) Cerebral arteriovenous oxygen difference (AVDO$_2$) during stepwise and abrupt increase of the concentration of isoflurane for induction of deliberate hypotension. Acta Anaesthesiol Scand [Suppl 91]: 142

Harp JR, Nilsson L, Siesjö BK (1976) The effect of halothane upon cerebral oxygen consumption in the rat. Acta Anaesth Scand 20: 83–90

Harris CE, Murray AM, Anderson JM, Grounds RM, Morgan M (1988) Effects of thiopentone, etomidate and propofol on the haemodynamic response to tracheal intubation. Anaesthesia 43 [Suppl]: 32–36

Harrison JL (1986) Postoperative seizures after isoflurane anesthesia. Anesth Analg 65: 1235–1236

Hartmann A, Buttinger C, Rommel T, Czernicki Z, Trtinjiak F (1989) Alteration of intracranial pressure, cerebral blood flow, autoregulation and carbondioxide-reactivity by hypotensive agents in baboons with intracranial hypertension. Neurochirurgica 32: 37–43

Hartung J, Cottrell JE (1987) Nitrous oxide reduces thiopental-induced prolongation of survival in hypoxic and anoxic mice. Anesth Analg 66: 47–52

Hartung HJ (1987) Beeinflussung des intrakraniellen druckes durch propofol (disoprivan). Anaesthesist 36: 66–68

Hartung HJ (1987) Intrakranielles Druckverhalten bei Patienten mit Schädel-Hirn-Trama nach Propofol- bzw Thiopental-Applikation. Anaesthesist 36: 285–287

Henderson JJ, McGeorg A, Teasdale GM (1982) Pharmacodynamics of althesin infusion: Electroencephalographic studies. Sixth European Congress of Anaesthesiology, London. Anaesthesia 143

Henneghan CPH, Thornton C, Navaratnarajah M, Jones JG (1987) Effect of isoflurane on auditory evoked response in man. Br J Anaesth 59: 277–282

Henriksen HT, Balslev Jørgensen P (1973) The effect of nitrous oxide on intracranial pressure in patients with intracranial disorders. Br J Anaesth 45: 486–492

Henriksen L, Thorshauge C, Harmsen A, Christensen P, Sørensen MB, Lester J,

Paulson OB (1983) Controlled hypotension with sodium nitroprusside: Effects on cerebral blood flow and cerebral venous blood gases in patients operated for cerebral aneurysms. Acta Anaesthesiol Scand 27: 62–67

Henriksen L (1986) Brain luxury perfusion during cardiopulmonary bypass in humans. A study of the cerebral blood flow response to changes in CO_2, O_2, and blood pressure. J Cereb Blood Flow Metab 6: 366–378

Herkenham M (1981) Anesthetics and the habenulo-interpeduncular system: selective sparing of metabolic activity. Brain Res 210: 461–466

Herrschaft H, Schmidt H, Gleim F, Albus G (1975) The response of human cerebral blood flow to anesthesia with thiopentone, methohexitone, propanidide, ketamine and etomidate. In: Penzholz H, Brock M et al (eds) Advances in neurosurgery, Springer, Berlin Heidelberg New York, pp 120–133

Hewer AJH, Logue V (1962) Methods of increasing the safety of neuroanaesthesia in the sitting position. Anaesthesia 17: 476–482

Hickey R, Bunegin L, Albin MS et al (1986) Cerebral blood flow responses during varying rates of isoflurane induced hypotension. Anesthesiology 65: A580

Hickey R, Albin MS, Bunegin L et al (1986) Intracranial pressure dynamics during and subsequent to induced hypotension with isoflurane. Anesthesiology 65: A460

Hickey R, Albin MS, Bunegin L (1989) Ketamine abolishes central nervous autoregulation in the rat. Anesth Analg 68: S120

Hicks RG, Kerr DR, Horton DA (1986) Thiopentone cerebral protection under EEG control during carotid endarterectomy. Anaesth Intens Care 14: 22–28

Higgins JR, Strunk BL, Schiller NB (1984) Diagnosis of paradoxical embolism with contrast echocardiography. Am Heart J 107: 375–377

Hilfiker O, Kettler D (1981) Die Wirkung von Midazolam auf die Hirndurchblutung beim Tier und beim Mensch. Arzneim Forsch 31: 2236–2237

Hodes JE, Soncrant TT, Larson DM, Carlson SG, Rapoport SI (1983) Selective changes in local cerebral glucose utilization induced by phenobarbital in the rat. Anesthesiology 63: 633–639

Hodges MR, Stanley TH, Johansen RK (1977) Pulmonary shunt and cardiovascular responses to CPAP during nitroprusside-induced hypotension. Anesthesiology 46: 339–341

Hoff J, Schmith A, Nielsen S, Larson P (1973) Effects of barbiturate and halothane anaesthesia on focal cerebral infarction in the dog. Surg Forum 24: 449–452

Hoff JT, Smith AL, Hankinson HL, Nielsen SL (1975) Barbiturate protection from cerebral infarction in primates. Stroke 6: 28–33

— Pitts LH, Spetzler R, Wilson CB (1977) Barbiturates for protection from cerebral ischaemia in aneurysm surgery. Acta Neurol Scand 56: 158–159

Hoffman WE, Miletich DJ, Albrecht RF (1982a) Cardiovascular and regional blood flow changes during halothane anesthesia in the aged rat. Anesthesiology 56: 444–448

— Albrecht RF, Miletich DJ (1982b) Cerebrovascular and metabolic effects of SNP-induced hypotension in young and aged hypertensive rats. Anesthesiology 56: 427–430

— Miletich DJ, Albrecht RF (1986) The effects of midazolam on cerebral blood flow and oxygen consumption and its interaction with nitrous oxide. Anesth Analg 65: 729–733

Hoka S, Siker D, Bosnjak ZJ, Kampine JP (1987) Alteration of blood distribution and vascular capacitance during induced hypotension in deafferented dogs. Anesthesiology 66: 647–652

Homi J, Konchigeri HN, Exkenhoff JE, Linde HW (1972) A new anesthetic agent—Forane. Preliminary observations in man. Anesth Analg 51: 439–477

Hosobuchi Y, Baskin DS, Woo SK (1982) Reversal of neurological deficits by opiate antagonist nalaxone after cerebral ischaemia in animals and humans. J Cereb Blood Flow Metab [Suppl 1]: 98–100

Hossmann KA (1976) Development and resolution of ischaemic brain swelling. In: Pappius HM, Feindal W (eds) Dynamics of brain edema. Springer, Berlin Heidelberg New York, pp 219–228

Hougaard K, Hansen A, Brodersen P (1974) The effect of ketamine on regional cerebral blood flow in man. Anesthesiology 41: 562–567

Huang M, Drummond GI (1979) Adenylate cyclase in cerebral microvessels: Action of guanine nucleotides, adenosine and other agonists. Mol Pharmacol 16: 462–472

Hunter AR (1972) Thiopentone supplemented anaesthesia for neurosurgery. Br J Anaesth 44: 506–510

Hymes JA (1985) Seizure activity during isoflurane anesthesia. Anesth Analg 64: 367–368

Høedt-Rasmussen K (1967) Regional cerebral blood flow. The intra-arterial injection method (thesis). Munksgaard, Kopenhagen, pp 1–81

Iannotti F, Hoff J (1983) Ischemic brain edema with and without reperfusion: An experimental study in gerbils. Stroke 14: 562–567

Inada E, Cullen DJ, Nemeskal R, Teplick R (1987) Effect of labetamol on the hemodynamic response to intubation: A controlled randomized double-blind study. Anesthesiology 67: A31

Ingvar M, Abdul-Rahman A, Siesjö BK (1980) Local cerebral glucose consumption in the artificially ventilated rat: influence of nitrous oxide analgesia and of phenobarbital anesthesia. Acta Physiol Scand 109: 177–185

— Shapiro HM (1981) Selective metabolic activation of the hippocampus during lidocaine-induced pre-seizure activity. Anesthesiology 54: 33–37

— Siesjö BK (1982) Effect of nitrous oxide on local cerebral glucose utilization in rats. J Cereb Blood Flow Metab 2: 481–486

Ishikawa T, McDowall DG (1980) Electrical activity of the cerebral cortex during induced hypotension with sodium nitroprusside and trimethaphan in the cat. Br J Anaesth 53: 605–611

— Funatsu N, Okamoto K, Takeshita H, McDowall DG (1983) Blood–brain barrier function following drug-induced hypotension in the dog. Anesthesiology 59: 526–531

Ivankovich AD, Miletich DJ, Albrecht RF, Zahed B (1976) Sodium nitroprusside and cerebral blood flow in anesthetized and unanesthetized goat. Anesthesiology 44: 21–26

— Braverman B, Shulman M, Klowden AJ (1982) Prevention of nitroprusside toxicity with thiosulfate in dogs. Anesth Analg 61: 120–126

Iwatsuki N, Kuroda N, Amaha K, Iwatsuki K (1980) Succinylcholine-induced hyperkalemia in patients with ruptured cerebral aneurysms. Anesthesiology 53: 64–67

Jack RD (1974) Toxicity of sodium nitroprusside. Br J Anaesth 46: 952

Jafar JJ, Johns LM, Mullian SF (1986) The effect of mannitol on cerebral blood flow. J Neurosurg 64: 754–759

James HE (1978) Cytotoxic edema produced by 6-Aminonicotinamide and its response to therapy. Neurosurgery 3: 196–200

— (1980) Methodology for the control of intracranial pressure with hypertonic mannitol. Acta Neurochir (Wien) 51: 161–172

Jennett WB, McDowall DG, Barker J (1967) The effect of halothane on intracranial pressure in cerebral tumors. J Neurosurg 26: 270–274

— Barker J, Fitch W, McDowall DG (1969) Effect of anesthesia on intracranial pressure in patients with space-occupying lesions. Lancet i: 61–64

Jennett WB, Teasdale C (1981) Management of head injuries in the acute state. FA Davis Company, Philadelphia, pp 240–241

Jensen KA, Bünemann L, Øhrstrøm J, Nissen I, Hease J (1989) CBF in the sitting versus the supine position. J Cereb Blood Flow Metab (Suppl 1): S613

Joas TA, Stevens WC, Weger II EI (1971) Electroencephalographic seizure activity in dogs during anesthesia. Br J Anaesth 43: 739–745

Jobes DR, Kennell E, Bitner R, Swenson E, Wollman H (1975) Effects of morphine-nitrous oxide anesthesia on cerebral autoregulation. Anesthesiology 42: 30–34

— — Buch GL, Mull TD, Lecky JH, Behar MG, Wollman H (1977) Cerebral blood flow and metabolism during morphine-nitrous oxide anesthesia in man. Anesthesiology 47: 16–18

Johnstone IH, Harper AM (1973) The effect of mannitol on cerebral blood flow. An experimental study. J Neurosurg 38: 461–471

Jones JV, Fitch W, MacKenzie ET, Strandgaard S, Harper AM (1976) Lower limit of cerebral blood flow autoregulation in the baboon. Circ Res 39: 555–557

Julien RM, Kavan EM (1972) Electrographic studies of a new volatile anaesthetic agent: Enflurane (Ethrane). J Pharmacol Exper Therapeut 183: 393–403

Kaieda R, Todd MM, Cook LN, Warner DS (1989a) The effects of anesthetics and $PaCO_2$ on the cerebrovascular, metabolic, and electroencephalographic responses to nitrous oxide in the rabbit. Anesth Analg 68: 135–143

— — — — (1989b) Acute effects of changing plasma osmolality and colloid oncotic pressure on the formation of brain edema after cryogenic injury. Neurosurgery 24: 671–678

Karlin A, Hartung J, Cottrell JE (1988) Rate of induction of hypotension with trimethaphan modifies the intracranial pressure response in cats. Br J Anaesth 60: 161–166

Kassell NF, Hitchon PW, Gerk MK, Sokoll MD, Hill TR (1980) Alterations in cerebral blood flow, oxygen metabolism, and electrical activity produced by high dose sodium thiopental. Neurosurgery 7: 598–603

Kassell NF, Hitchon PW, Gerk MK, Sokoll MD, Hill TR (1981) Influence of changes in arterial pCO_2 on cerebral blood flow and metabolism during high-dose barbiturate therapy in dogs. J Neurosurg 54: 615–619

— Baumann KW, Hitchon PW, Gerk MK, Hill TR, Sokoll MD (1982) The effects of high dose mannitol on cerebral blood flow in dogs with normal intracranial pressure. Stroke 13: 59–61

— Boarini DJ, Olin JJ, Sprowell JA (1983a) Cerebral and systemic circulatory effects of arterial hypotension induced by adenosine. J Neurosurg 58: 69–76

— — Sprowell JA, Olin JJ (1983b) Pharmacologically induced profound arterial hypotension in the anesthetized dog. J Neurosurg 58: 77–83

Katsuki S, Arnold W, Murad F (1977) Effects of sodium nitroprusside, nitroglycerin, and sodium azide on levels of cyclic nucleotides and mechanical activity of various tissues. J Cyclic Nucleotides Res 3: 239–246

Katz J, Luehr S, Hills BA, Butler BD (1985) Halothane decreases pulmonary filtration of venous air emboli. Anesthesiology 63: A523

Katz JJ, Todd MM, Warner DS (1988) Quantitative comparison of cerebral blood volume in rats receiving halothane or isoflurane. Anesthesiology 69: A534

Katz JM, Abou-Madi M, Abou-Madi N, Trop D (1986) Do mannitol-induced haemodynamic responses influence its effect on intracranial pressure? A study in the dog with and without induced intracranial hypertension. Canad Anaesth Soc J 33: S81–S82

Kayama Y, Iwama K (1972) The EEG, evoked potentials, and single-unit activity during ketamine anaesthesia in cats. Anesthesiology 36: 316–328

— (1974) Comparison of the effects of althesin and barbiturates on the neocortex and hippocampus in cats. Br J Anaesth 46: 912–917

Keaney NP, McDowall DG, Pickerodt VWA, Turner JM, Lane JR, Okuda Y, Deshmukh VD, Coroneos NJ (1978) Time course of the cerebral circulatory response to metabolic depression. Am J Physiol 234: H74–H79

Keats AS (1985) Seizures from isoflurane. Anesth Analg 64: 1225–1224

Kety SS, Schmidt CF (1945) The determination of cerebral blood flow in man by the use of nitrous oxide in low concentrations. Am J Physiol 143: 53–66

— — (1948) The nitrous oxide method for the quantitative determination of cerebral blood flow in man: Theory, procedure and normal values. J Clin Lab Invest 27: 476–483

Keykhah MM, Smith DS, Carlsson C, Sato Y, Englebach I, Harp JR (1985) Influence of sufentanil on cerebral metabolism and circulation in the rat. Anesthesiology 63: 274–277

— — O'Neil JJ, Englebach I, Carson J, Harp JR (1986) The influence of etomidate on brain metabolites during severe hypoxia. Anesthesiology 65: A251

— — — Harp JR (1988) The influence of fentanyl upon cerebral high-energy metabolites, lactate, and glucose during severe hypoxia in the rat. Anesthesiology 69: 566–570

Khambatta HJ, Stone JG, Khan E (1979) Hypertension during anesthesia on discontinuation of sodium nitroprusside-induced hypotension. Anesthesiology 51: 127–130

— — — (1982) Captopril attenuates vasoactive hormonal release during nitro-prusside-induced hypotension. Anesthesiology 57: A62

— — Matteo RS, Khan E (1984) Propranolol premedication blunts stress response to nitroprusside hypotension. Anesth Analg 63: 125–128

Kien ND, White DA, Reitan JA, Eisele JH (1987) Cardiovascular function during controlled hypotension induced by adenosine triphosphate or sodium nitroprusside in the anesthetized dog. Anesth Analg 66: 103–110

Kiersey DK, Bickford RD, Faulconer A (1951) Electroencephalographic patterns produced by thiopental sodium during surgical operations: Description and classification. Br J Anaesth 23: 141–152

Kitahata LM, Gali ich JM, Sato I (1971) The effect of passive hyperventilation on intracranial pressure. J Neurosurg 34: 185–193

Klausen NO, Moesgaard J, Ferguson AH, Kaalund Jensen J, Larsen C, Paaby P (1983) Negative synacthen test during etomidate infusion. Lancet ii: 848

Klose R, Hartung HJ, Kotsch R, Walz T (1982) Experimentelle Untersuchungen zur intracraniellen Drucksteigerung durch Ketamine beim haemorrhagischen Schock. Anaesthesist 31: 33–38

Knight PR, Lane GA, Hensinger RN, Bolles RS, Bjoraker DG (1983) Catecholamine and renin-angiotensin response during hypotensive anesthesia induced by sodium nitroprusside or trimethaphan camsylate. Anesthesiology 59: 248–253

Knudsen L, Cold GE, Jensen S, Holdgaard HO, Johansen UT, Jensen S (1989) Cerebral blood flow (CBF) and metabolism (CMRO$_2$) during midazolam infusion. Interaction with nitrous oxide, and reversal with flumazenil in patients subjected to craniotomy for supratentorial cerebral tumours. Acta Anaesthesiol Scand [Suppl] 91: 33: 142

Kobrine AI, Evans DE, Le Grys DC et al (1984) Effects of intravenous lidocaine on experimental spinal cord injury. J Neurosurg 60: 595–601

Koch KA, Jackson DL, Schmiedl M, Thompson WL, Rosenblatt JI (1984) Effect of thiopental therapy on cerebral blood flow after total cerebral ischaemia. Crit Care Med 12: 90–95

Kochs E, Roewer N, Peter A, Schulte am Esch J (1988) Wirkungen von Flumazenil auf den globalen zerebralen Blutfluss und den intrakraniellen Druck in der Reperfusionsphase nach globaler inkompletter zerebraler Ischäemie. Anaesth Intensivther Notfallmed 23: 159–162

Kofke WA, Snider MT, Young RSK, Ramer JC (1985) Prolonged low flow isoflurane anesthesia for status epilepticus. Anesthesiology 62: 653–656

— Hawkins RA, Davis DW, Biebuyck JF (1986) Increased brain glucose during isoflurane anesthesia. Anesthesiology 65: A582

Kolassa N, Pfleger K (1975) Adenosine uptake by erythrocytes of man, rat, and guinea-pig and its inhibition by hexobendine and dipyridamole. Biochem Pharmacol 24: 154–156

Kontos HA, Wei EP (1986) Superoxide production in experimental brain injury. J Neurosurg 64: 803–807

Kraynack BJ, Gintautas J, Hinshaw J (1981) Adenosine triphosphate increases survival time during hyoxia. Neuropharm 20: 887–890

Krieger W, Copperman J, Laxer KD (1985) Seizures with etomidate anesthesia. Anesth Analg 64: 1226–1227

Kreuscher H (1967) Die Hirndurchblutung unter Neuroleptanaesthesie. Springer, Berlin Heidelberg New York

Krieglstein J, Sperling G, Twietmeyer G (1981) Effects of thiopental on regulatory mechanisms of brain energy metabolism. Arch Pharmacol 318: 56–61

Kruczek M, Albin MS, Wolf S, Bertoni JM (1980) Postoperative seizure activity following enflurane anesthesia. Anesthesiology 53: 175–176

Kuramoto T, Oshita S, Takeshita H, Ishikawa T (1979) Modification of the relationship between cerebral metabolism, blood flow, and electroencephalogram by stimulation during anaesthesia in the dog. Anesthesiology 51: 211–217

Lafferty JJ, Keykhah MM, Shapiro HM, VanHorn K, Behar MG (1978) Cerebral hypometabolism obtained with deep pentobarbital anesthesia and hypothermia (30 C). Anesthesiology 49: 159–164

Lagerkranser M, Irestedt L, Sollevi A, Andreen M (1984) Central and splanchnic hemodynamics in the dog during controlled hypotension with adenosine. Anesthesiology 60: 547–552

Lagerkranser M, Sollevi A, Irestedt L, Tidgren B, Andreen M (1985) Renin release during controlled hypotension with sodium nitroprusside, nitroglycerin and adenosine: A comparative study in the dog. Acta Anaesthesiol Scand 29: 45–49

— Bergstrand G, Gordon E, Irestedt L, Lindquist C, Stånge K, Sollevi A (1989) Cerebral blood flow and metabolism during adenosine-induced hypotension in patients undergoing cerebral aneurysm surgery. Acta Anaesthesiol Scand 33: 15–20

Lam AM, Gelb AW (1983) Cardiovascular effects of isoflurane-induced hypotension for cerebral aneurysm surgery. Anesth Analg 62: 742–748

— Wu X, Gelb AW (1988) Regional cerebral blood flow during nicardipine and nitroprusside induced hypotension. Anesth Analg 67: S125

Lanier WL, Milde JH, Michenfelder JD (1985) The cerebral effects of pancuronium and atracurium in halothane-anesthetized dogs. Anesthesiology 63: 589–597

— — — (1986) Cerebral stimulation following succinylcholine in dogs. Anesthesiology 64: 551–559

Larsen R, Hilfiker O, Radle J, Sonntag H (1981) Midazolam: Wirkung auf allgemeine Häemodynamik, Hirndurchblutung und cerebralen Sauerstoffverbrauch bei neurochirurgischen Patienten. Anaesthesist 30: 18–21

— Teichmann J, Hilfiker O, Busse C, Sonntag H (1982) Nitroprusside-hypotension: Cerebral blood flow and cerebral oxygen consumption in neurosurgical patients. Acta Anaesth Scand 26: 327–330

— Maurer I, Khambatta H (1988) Wirkungen von Isofluran und Enfluran auf die zerebrale Häemodynamik und den zerebralen Sauerstoffverbrauch des Menschen. Anesthesist 37: 173–181

Lassen NA, Munck O (1955) The cerebral blood flow in man determined by the use of radioactive krypton. Acta Physiol Scand 33: 30–49

— Lane MH (1961) Validity of internal jugular blood for study of cerebral blood flow and metabolism. J Appl Physiol 16: 313–320

— Klee A (1965) Cerebral blood flow determined by saturation and desaturation with Krypton-85. An evaluation of the validity of the inert gas method of Kety and Schmidt. Circ Res 16: 26–32

— Christensen MS (1976) Physiology of cerebral blood flow. Br J Anaesth 48: 719–735

Laurent JP, Lawner P, Simeone FA, Fink E (1982) Pentobarbital changes compartmental contribution to cerebral blood flow. J Neurosurg 56: 504–510

Laurito CE, Baugman VL, Polek WV, Riegler FX, VadeBoncouer TR (1987) Aerosolized and intravenous lidocaine are no more effective than placebo for the control of hemodynamic response to intubation. Anesthesiology 67: A29

Lawner PM, Simeone FA (1979) Treatment of intraoperative middle cerebral artery occlusion with pentobarbital and extracranial–intracranial bypass. J Neurosurg 51: 710–712

Laycock JRD, Coakham HB, Silver IA, Walters FJM (1986) Changes in brain surface oxygen tension during profound hypotension induced with sodium nitroprusside or adenosine in the sheep. Br J Anaesth 58: 1422–1426

Lebowitz MH, Blitt CD, Dillon JB (1972) Enflurane-induced central nervous system excitation and its relation to carbon dioxide tension. Anesth Analg 51: 355–363

Ledingham IM, Watt I (1983) Influence of sedation on mortality in critically ill multiple trauma patients. Lancet i: 1270

Leech PJ, Miller JD, Fitch W, Barker J (1974) Cerebral blood flow, internal carotid artery pressure, and EEG as a guide to the safety of carotid ligation. J Neurol Neurosurg Psychiatry 37: 854–862

Leivers D, Spilsbury RA, Young JVI (1971) Air embolism during neurosurgery in the sitting position. Br J Anaesth 43: 84–90

Leslie JB, Kalayjian RW, McLoughlin TM, Russell RD, Plachetka JR (1987) Ablation of the hemodynamic responses to tracheal intubation with preinduction intravenous labetalol. Anesthesiology 67: A30

Levy DE, Brierley JB (1979) Delayed pentobarbital administration limits ischemic brain damage in gerbils. Ann Neurol 5: 59–64

Lightfoote WE, Molinari GF, Chase TN (1977) Modification of cerebral ischemic damage by anesthetics. Stroke 8: 627–628

Lin DM, Powell HC, Shapiro HM (1986) Absence of longterm neuropathology after sustained enflurane epileptiform activity. Anesthesiology 65: A348

List WF, Crumrine RS, Cascorbi HF, Weiss F (1972) Increased cerebrospinal fluid pressure after ketamine (correspondence). Anesthesiology 36: 98

Litt L, Gonzales-Mendez R, James TL, Sessler DI, Mills P, Chew W, Moseley M, Pereira B, Severinghaus JW, Hamilton WK (1987) An in vivo study of halothane uptake and elimination in the rat brain with fluorine nuclear magnetic resonance spectroscopy. Anesthesiology 67: 161–168

Little JR (1978) Modification of acute focal ischemia by treatment with mannitol. Stroke 9: 4–9

Little JR, Latchaw JP, Slugg RM, Lesser RP, Stowe NT (1982) Treatment of acute focal cerebral ischemia with propranolol. Stroke 13: 302–307

Lobe DP, Brauer FS (1983) Barbiturate protection in extracranial–intracranial anastomosis? Anesthesiology 59: A331

Long DM, Maxwell R, Choi KS (1976) A new therapy regimen for brain edema. In: Pappius HM, Feidal W (eds) Dynamics of brain edema. Springer, Berlin Heidelberg New York, pp 293–300

Longnecker DE (1984) Stress free: To be or not to be? (editorial). Anesthesiology 61: 643–644

— Seyde WC (1984) Cerebral oxygen tension during deliberate hypotension with sodium nitroprusside, 2-Cl-adenosine, or deep isoflurane anesthesia in rats. Anesthesiology 61: A366

Lu GP, Chi OZ, Kaul DK, Baez S, Orkin LR (1982) Effects of nitroprusside on cerebral microcirculation. Anesthesiology 57: A49

— Prinscott J, Frost EAM, Gibson JA, Goldiner PL (1987) Different response of bolus or continuous infusion of sufentanil on brain microcirculation. Anesthesiology 67: A89

Lüben V, Hempelmann G (1982) Improved deep controlled hypotension in aneurysmal surgery. Acta Neurochir (Wien) 60: 201–214

Madsen JB, Cold GE, Eriksen HO, Eskesen V, Blatt-Lyon B (1986) CBF and $CMRO_2$ during craniotomy for small supratentorial cerebral tumours in enflurane anaesthesia. A dose-response study. Acta Anaesthesiol Scand 30: 633–636

— — Hansen ES, Bardrum B (1987a) Cerebral blood flow, cerebral metabolic rate of oxygen and relative CO_2 reactivity during craniotomy for supratentorial cerebral tumours in halothane anaesthesia. A dose-response study. Acta Anaesthesiol Scand 31: 454–457

— — (1987) Cerebral blood flow and oxygen uptake during anesthesia with halothane, enflurane or isoflurane. J Cereb Blood Flow Metab 7 [Suppl 1]: S628

— — Hansen ES, Bardrum B (1987b) The effect of isoflurane on cerebral blood flow and metabolism in humans during craniotomy for small supratentorial cerebral tumors. Anesthesiology 66: 332–336

— — — — Kruse-Larsen C (1987c) Cerebral blood flow and metabolism during isoflurane-induced hypotension in patients subjected to surgery for cerebral aneurysms. Br J Anaesth 59: 1204–1207

— Guldager H, Jensen FM (1989) CBF and $CMRO_2$ during neuroanaesthesia with continuous infusion of propofol. Acta Anaesthesiol Scand [Suppl 91] 33: 143

Maekawa T, Sakabe T, Takeshita H (1974) Diazepam blocks cerebral metabolic and circulatory responses to local anesthetic-induced seizures. Anesthesiology 41: 389–391

Maekawa T, McDowall DG, Okuda Y (1979) Brain surface oxygen tension and cerebral cortical blood flow during hemorrhagic and drug-induced hypotension in the cat. Anesthesiology 51: 313–320

— Oshibuchi T, Takeshita H, Imamura A (1981) Cerebral energy state and glycolytic metabolism during lidocaine infusion in the rat. Anesthesiology 54: 278–283

— Tommasino C, Shapiro MD, Keifer J (1983) Local cerebral blood flow during isoflurane anesthesia. Anesthesiology 59: A308

— — Shapiro HM, Keifer-Goodman J, Kohlenberger RW (1986) Local cerebral blood flow and glucose utilization during isoflurane anesthesia in the rat. Anesthesiology 65: 144–151

Maktabi M, Warner D, Boarini D, Sokoll M, Adolphson A (1985) A comparison of nitroprusside, nitroglycerin and deep isoflurane anesthesia for induced hypotension. Anesthesiology 63: A403

— Warner DS, Sokoll MD (1986) The effects of hypotension induced by sodium nitroprusside and isoflurane on serum catecholamines and plasma renin activity. Anesthesiology 65: A578

Mann JD, Cookson SL, Mann ES (1980) Differential effects of pentobarbital, ketamine hydrochloride, and enflurane anesthesia on CSF formation rate and outflow resistance in the rat. In: Shulman K, Marmarou A, Miller JD, Becker DP, Hochwald GM, Brock M (eds) Intracranial pressure IV. Springer, Berlin Heidelberg New York, pp 466–471

Manninen PH, Lam AM, Gelb AW, Brown SC (1987) The effect of high-dose mannitol on serum and urine electrolytes and osmolality in neurosurgical patients. Can J Anaesth 34: 442–446

— Mahendran B, Gelb AW, Merchant R (1989) The effect of succinylcholine on serum potassium in patients with acutely ruptured cerebral aneurysms. Anesth Analg 68: S180

Manohar M, Parks C (1984) Regional distribution of brain and myocardial perfusion in swine while awake and during 1.0 and 1.5 MAC isoflurane anesthesia produced without and with 50% nitrous oxide. Cardiovasc Res 18: 344–353

March ML, Shapiro HM, Smith RW, Marshall LF (1979a) Changes in neurologic status and intracranial pressure associated with sodium nitroprusside administration. Anesthesiology 51: 336–338

— Aidinis SJ, Naughton KVH, Marshall LF, Shapiro HM (1979b) The technique of nitroprusside administration modifies the intracranial pressure response. Anesthesiology 51: 538–541

March ML, Dunlop BJ, Shapiro HM, Gagnon RL, Rockoff MA (1980) Succinylcholine-intracranial pressure effects in neurosurgical patients. Anesth Analg 59: 550–551

Marshall LF, Smith RW, Rauscher LA, Shapiro HM (1978) Mannitol dose requirements in brain-injured patients. J Neurosurg 48: 169–172

— Sang UH (1983) Treatment of massive intraoperative brain swelling. Neurosurgery 13: 412–414

Marshall BM (1965) Air embolism in neurosurgical anaesthesia, its diagnosis and treatment. Can Anaesth Soc J 12: 255–261

Marshall BM (1973) Neurolept anesthesia in neurosurgery. In: Oyama T (ed) International anesthesiology clinics. Little Brown and Company, pp 103–125

Marshall WK, Page RB, Milchak MA (1982) Furosemide reduces brain water in cerebral injury in dogs. Anesthesiology 57: A308

— Bedford RF, Miller ED (1983) Cardiovascular responses in the seated position—Impact of four anesthetic techniques. Anesth Analg 62: 648–653

Martin JT (1978) Positioning in anesthesia and surgery. Saunders Company, pp 44–79

Martin RW, Colley PS (1983) Evaluation of transesophageal doppler detection of air embolism in dogs. Anesthesiology 58: 117–123

Martins AN, Doyle TF, Wright SJ Jr, Bass BG (1980) Response of cerebral circulation to topical histamine. Stroke 11: 469–476

Marx GF, Andrews IC, Orkin LR (1962) Cerebrospinal fluid pressures during halothane anaesthesia. Can Anaesth Soc J 9: 239–245

Marx W, Shah N, Long C et al (1988) Sufentanil, alfentanil and fentanyl: Impact on CSF pressure in patients with brain tumors. Anesthesiology 69: A627

Marx W, Shah N, Long C, Arbit E, Galicich J, Mascott C, Mallya K, Bedford R (1989) Sufentanil, alfentanil, and fentanyl: Impact on cerebrospinal fluid pressure in patients with brain tumors. J Neurosurg Anesthesiol 1: 3–7

Matjasko J, Petrozza P, Cohen M, Steinberg P (1985) Anesthesia and surgery in the seated position: Analysis of 554 cases. Neurosurgery 17: 695–702

Matjasko MJ, Hellman J, MacKenzie CF (1987) Venous air embolism, hypotension, and end-tidal nitrogen. Neurosurgery 21: 378–382

Mazzarella B, Mastronardi P, Cafiero T, Gariulo G, Frangiosa A, Tomassino C, Stella L, De Chiara A (1986) Isoflurane and intracranial pressure. In: Miller JD, Teasdale GM, Rowan JO, Galbraith SL, Mendelow AD (eds) Intracranial pressure VI. Springer, Berlin Heidelberg New York, pp 732–735

Mazzoni P, Giffin JP, Cottrell JE, Hartung J, Capuano C, Epstein JM (1985) Intracranial pressure during diltiazem-induced hypotension in anesthetized dogs. Anesth Analg 64: 1001–1004

McDowall DG, Harper AM, Jacobson I (1963) Cerebral blood flow during halothane anaesthesia. Br J Anaesth 35: 394–402

— Heuser D, Okuda Y, Jones GM, Wadon A (1979) Relationship between cerebral blood flow changes and cortical extracellular fluid pH during cerebral metabolic depression induced by althesin. Br J Anaesth 51: 1109–1115

McDowall GD (1985) Induced hypotension and brain ischaemia. Br J Anaesth 57: 110–119

McHenry LC (1964) Quantitative cerebral blood flow determination. Application of a krypton 85 desaturation technique in man. Neurology (Minneap) 14: 785–793

— Slocum HC, Bivens HE, Mayes HA, Hayes GJ (1965) Hyperventilation in awake and anesthetized man. Arch Neurol 12: 270–277

McKay RD, Sundt TM, Michenfelder JD, Gronert GA, Messick JM, Sharbrough FW, Piepgras DG (1976) Internal carotid artery stump pressure and cerebral blood flow during carotid endarterectomy: Modification by halothane, enflurane, and innovar. Anesthesiology 45: 390–399

McKrell TN, Stone HH, Wechsler RL (1955) Effect of drug induced hypotension on the cerebral circulation in man. Surg Forum 5: 730–736

McLeskey CH, Cullen BF, Kennery RD, Galindo A (1974) Control of cerebral perfusion pressure during induction of anesthesia in high-risk neurosurgical patients. Anesth Analg 53: 985–992

McPherson RW, Johnson RM, Traystman RJ (1982) The effects of alfentanil on the cerebral vasculature. Anesthesiology 57: A354

— Traystman RJ (1984) Fentanyl and cerebral vascular responsivity in dogs. Anesthesiology 60: 180–186

— Traystman RJ (1987) Effects of isoflurane on cerebral autoregulation. Anesthesiology 67: A576

— Brian JE, Traystman RJ (1989) Cerebrovascular responsiveness to carbon dioxide in dogs with 1.4% and 2.8% isoflurane. Anesthesiology 70: 843–850

McQueen JD, Jeanes LD (1964) Dehydration and rehydration of the brain with hypertonic urea and mannitol. J Neurosurg 11: 118–128

Mendelow AD, Teasdale GM, Russell T, Flood J, Patterson J, Murray GD (1985) Effect of mannitol on cerebral blood flow and cerebral perfusion pressure in human head injury. J Neurosurg 63: 43–48

Mercier J, Mercier E (1955) Action de quelques alcaloides secondaires de l'opium sur l'electrocorticogramme du chien. C R Soc Biol 149: 760

Merckx L, Van Hemelrijck J, Van Aken H, Plets C, Goffin J (1988) Total intravenous anaesthesia using propofol and alfentanil infusion in neurosurgical patients. Anesthesiology 69: A576

Merrifield AJ, Blundell MD (1974) Toxicity of sodium nitroprusside. Br J Anaesth 46: 324

Messick JM, Theye RA (1969) Effects of pentobarbital and meperidine on canine cerebral and total oxygen consumption rates. Can Anaesth Soc J 16: 321–330

— Newberg LA, Nugent M, Faust RJ (1985) Principles of neuroanesthesia for the nonneurosurgical patient with CNS pathophysiology. Anesth Analg 64: 143–174

— Casement B, Sharbrough FW, Milde LN, Michenfelder JD, Sundt TM (1987) Correlation of regional cerebral blood flow (rCBF) with EEG changes during isoflurane anesthesia for carotid endarterectomy: Critical rCBF. Anesthesiology 66: 344–349

Mehta M, Sokoll MD (1981) Relation of right and left atrial pressure during venous air embolism. Anesthesiology 55: A238

Meyer FB, Anderson RE, Sundt TM, Yakch TL (1987) Treatment of experimental focal cerebral ischemia with mannitol. J Neurosurg 66: 109–115

Michenfelder JD, Martins JT, Altenburg BM, Rehder K (1969) Air embolism during neurosurgery: an evaluation of right-atrial catheters for diagnosis and treatment. J Am Med Ass 298: 1353–1358

— VanDyke RA, Theye RA (1970) The effect of anesthetic agents and techniques on canine cerebral ATP and lactate levels. Anesthesiology 33: 315–321

— Theye RA (1971) Effects of fentanyl, droperidol, and innovar on canine cerebral metabolism and blood flow. Br J Anaesth 43: 630–636

— Miller RH, Gronert GA (1972) Evaluation of an ultrasonic device (Doppler) for the diagnosis of venous air embolism. Anesthesiology 36: 164–167

Michenfelder JD, Theye RA (1973) Cerebral protection by thiopental during hypoxia. Anesthesiology 39: 510–517

— Cucchiara RF (1974) Canine cerebral oxygen consumption during enflurane anesthesia and its modification during induced seizures. Anesthesiology 40: 575–580

— (1974) The interdependency of cerebral functional and metabolic effects following massive doses of thiopental in the dog. Anesthesiology 41: 231–236

— Theye RA (1975) *In vivo* toxic effects of halothane on canine cerebral metabolic pathways. Am J Physiol 229: 1050–1055

— Milde JH, Sundt JM Jr (1976) Cerebral protection by barbiturate anesthesia. Use after middle cerebral artery occlusion in Java monkeys. Arch Neurol 33: 345–350

— Theye RA (1977) Canine systemic and cerebral effects of hypotension induced by hemorrhage, trimethaphan, halothane or nitroprusside. Anesthesiology 46: 188–195

— (1977) Cyanide release from sodium nitroprusside in the dog. Anesthesiology 46: 196–201

— (1986) A valid demonstration of barbiturate-induced brain protection in man—At last. Anesthesiology 64: 140–142

— (1987) Does isoflurane aggravate regional cerebral ischemia? (Editorial). Anesthesiology 66: 451–452

— Sundt TM, Fode N, Sharbrough FW (1987) Isoflurane when compared to enflurane and halothane decreases the frequency of cerebral ischaemia during carotid endarterectomy. Anesthesiology 67: 336–340

— Milde JH (1988) The interaction of sodium nitroprusside, hypotension, and isoflurane in determining cerebral vasculature effects. Anesthesiology 69: 870–875

Milde LN, Milde JH, Michenfelder JD (1985) Cerebral functional, metabolic, and hemodynamic effects of etomidate in dogs. Anesthesiology 63: 371–377

— — (1986) Preservation of cerebral metabolites by etomidate during incomplete cerebral ischemia in dogs. Anesthesiology 65: 272–277

— — (1987) The cerebral hemodynamic and metabolic effects of sufentanil in dogs. Anesthesiology 67: A570

— — (1987) The detrimental effect of lidocaine on cerebral metabolism measured in dogs anesthetized with isoflurane. Anesthesiology 67: 180–184

— — Lanier WL, Michenfelder JD (1988) Comparison of the effects of isoflurane and thiopental on neurologic outcome and neuropathology after temporary focal cerebral ischemia in primates. Anesthesiology 69: 905–913

— (1988) The hypoxic mouse model for screening cerebral protective agents: A re-examination. Anesth Analg 67: 917–922

— Milde JH (1989) Cerebral effects of sufentanil in dogs with reduced intracranial compliance. Anesth Analg 68: S196

Miletich DJ, Ivankovich AD, Albrecht RF, Reimann CR, Rosenberg R, McKissic ED (1976) Absence of autoregulation of cerebral blood flow during halothane and enflurane anesthesia. Anesth Analg 55: 100–109

Millar RA (1972) Neurosurgical anaesthesia in the sitting position. A report of experience with 110 patients using controlled or spontaneous ventilation. Br J Anaesth 44: 495–505

Miller CL, Lampard DG, Griffiths RI, Brown WA (1978) Local cerebral blood flow and the electrocorticogram during sodium nitroprusside hypotension. Anaesth Intensive Care 6: 290–296

Miller JD, Leech P (1975) Effects of mannitol and steroid therapy on intracranial volume–pressure relationships in patients. J Neurosurg 42: 274–281

— (1979) Barbiturates and raised intracranial pressure. Ann Neurol 6: 189–193

Miller R, Tausk HC, Stark DCC (1975) Effect of innovar, fentanyl and droperidol on the cerebrospinal fluid pressure in neurosurgical patients. Can Anaesth Soc J 22: 502–508

Mills P, Sessler DI, Moseley M, Chew W, Pereira B, James TL, Litt L (1987) An *in vivo* 19-F nuclear magnetic resonance study of isoflurane elimination from the rabbit brain. Anesthesiology 67: 169–173

Minton MD, Stirt JA, Bedford RF, Hawworth C (1985) Intracranial pressure after atracurium in neurosurgical patients. Anesth Analg 64: 1113–1116

— — — (1986a) Effect of succinylcholine on serum potassium in patients with brain tumors. Anesth Analg 65: S100

— — Bedford RF (1986b) Vecuronium and intracranial pressure in man. Anesth Analg 65: S101

— Grosslight K, Stirt JA, Bedford RF (1986c) Increases in intracranial pressure from succinylcholine: Prevention by prior nondepolarizing blockade. Anesthesiology 65: 165–169

Mirsky AF, Stockard JJ, Skoff BF, Jones TA (1979) Brainstem auditory evoked potential alterations during induced and spontaneous generalized spike-wave activity in animals and humans, Neuroscience 5: 196

Misfeldt BB, Balslev Jørgensen P, Rishøj M (1974) The effect of nitrous oxide and halothane upon the intracranial pressure in hypocapnic patients with intracranial disorders. Br J Anaesth 46: 853–858

Moffat JA, McDougall MJ, Brunet D, Saunders F, Shelley ES, Cervenko FW, Milne B (1983) Thiopental bolus during carotid endarterectomy—rational drug therapy? Can Anaesth Soc J 30: 615–622

Mollbegott LP, Flachburg MH, Karasic HL, Karlin BL (1987) Probable seizures after sufentanil. Anesth Analg 66: 91–93

Monk CR, Sperry RJ, Durieux ME, Walker MS, Longnecker DE (1987) The regional hemodynamic effects of induced hypotension with isoflurane, sodium nitroprusside or 2-chloroadenosine. Anesthesiology 67: A34

Möhler H, Okada T (1977) Benzodiazepines receptor: demonstration in the central nervous system. Science 198: 849–851

Mori K, Iwabuchi K, Fujito M (1973) The effects of depolarizing muscle relaxants on the electroencephalogram and the circulation during halothane anesthesia in man. Br J Anaesth 45: 604–610

Morii S, Nagai AC, Winn HR (1986) Reactivity of rat pial arterioles and venules to adenosine and carbon dioxide: With detailed description of the closed cranial window technique in rats. J Cereb Blood Flow Metab 6: 34–41

Morita H, Nemoto EM, Bleyaert AL, Stezoski SW (1977) Brain blood flow autoregulation and metabolism during halothane anesthesia in monkeys. Am J Physiol 233: H670–676

Morris PJ, Heuser D, McDowall DG, Hashiba M, Myers D (1983) Cerebral cortical extracellular fluid H^+ and K^+ activities during hypotension in cats. Anesthesiology 59: 10–18

Moss E, Powell D, Gibson RM, McDowall DG (1978) Effects of fentanyl on intracranial pressure and cerebral perfusion pressure during hypocapnia. Br J Anaesth 50: 779–784

— McDowall DG (1979) ICP increases with 50% nitrous oxide in oxygen in severe head injuries during controlled ventilation. Br J Anaesth 51: 757–760

— Powell D, Gibson RM, McDowall DG (1979) Effect of etomidate on intracranial pressure and cerebral perfusion pressure. Br J Anaesth 51: 347–351

— Dearden NM, McDowall DG (1983) Effects of 2% enflurane on intracranial pressure and cerebral perfusion pressure. Br J Anaesth 55: 1083–1087

Muizelaar JP, Wei EP, Kontos HA, Becker DP (1983) Mannitol causes compensatory cerebral vasoconstriction and vasodilatation in response to blood viscosity changes. J Neurosurg 59: 822–828

— Lutz HA, Becker DP (1984) Effect of mannitol on ICP and CBF and correlation with pressure autoregulation in severely head-injured patients. J Neurosurg 61: 700–706

Munari C, Casaroli D, Matteuzzi G, Pacifico L (1979) The use of althesin in drug-resistant status epilepticus. Epilepsia 20: 475–483

Munson ES (1971) Effect of nitrous oxide on the pulmonary circulation during venous air embolism. Anesth Analg 50: 785–792

Murkin JM, Farrar JK, Tweed WA, Guiraudon G, McKenzie FN (1985) Relationship between cerebral blood flow and O_2 consumption during high-dose narcotic anesthesia for cardiac surgery. Anesthesiology 63: A44

Murkin JM, Farrar JK, Tweed WA, Guiraudon G (1986) Cerebral blood flow, oxygen consumption and EEG during isoflurane anesthesia. Anesth Analg 65: S107

— — — (1988) Sufentanil anaesthesia reduces cerebral blood flow and cerebral oxygen consumption. Can J Anaesth 35: S131

— — (1989) The influence of high dose fentanyl–diazepam anesthesia on cerebral blood flow and cerebral oxygen consumption. Anesth Analg 68: S205

Murphy FL, Kennell EM, Johnstone RE et al (1974) The effects of enflurane, isoflurane and halothane on cerebral blood flow and metabolism in man. Abstracts and scientific papers. Annual meeting of the American Sos Anaesth 61–62

Murray IPC, Hoschl R, Choy D (1978) The jugular venous reflux. Clin Nucl Med 3: 56–57

Musella L, Wilder BJ, Schmidt RP (1971) Electroencephalographic activation with intravenous methohexital in psychomotor epilepsia. Neurology 21: 594–602

Mutch WAC, Ringaert KRA (1987) Effects of haemorrhage and phenylephedrine on cerebral blood flow in rats during isoflurane anaesthesia. Can J Anaesth S106–S107

Muzzi DA, Cucchiara RF, Oliver SB (1988) Labetalol and esmolol in the control of hypertension following intracranial surgery. Anesthesiology 69: A548

Myers RR, Shapiro HM (1979) Local cerebral metabolism during enflurane anesthesia: identification of epileptigenic foci. Electroenceph Clin Neurophysiol 47: 153–162

Nakamura K, Koide M, Imanaga T, Ogasawara H, Takahashi M, Yoshikawa M (1980) Prolonged neuromuscular blockade following trimethaphan infusion: A case report and in vitro study of cholinesterase inhibition. Anaesthesia 35: 1202–1207

Nakamura K, Hatano Y, Mori K (1988) The site of action of trimethaphan-induced neuromuscular blockade in isolated rat and frog muscle. Acta Anaesthesiol Scand 32: 125–130

Nath F, Galbraith S (1986) The effect of mannitol on cerebral white matter water content. J Neurosurg 65: 41–43

Nehls DG, Todd MM, Spetzler RF, Drummond JC, Thompson RA, Johnson PC (1987) A comparison of the cerebral protective effects of isoflurane and barbiturates during temporary focal ischaemia in primates. Anesthesiology 66: 453–464

Neigh JL, Garman JK, Harp JR (1971) The electroencephalographic pattern during anesthesia with ethrane: Effects of depth of anesthesia, PaCO$_2$ and nitrous oxide. Anesthesiology 35: 482–487

Nelson SR, Howard RB, Cross RS, Samson F (1980) Ketamine-induced changes in regional glucose utilization in the rat brain. Anesthesiology 52: 330–334

Neundörfer B, Klose R (1975) EEG-Veränderungen bei Kindern während Enflurane-Anästhesie. Prakt Anesth 10: 271–284

Newberg LA, Michenfelder JD (1983) Cerebral protection by isoflurane during hypoxemia or ischemia. Anesthesiology 59: 29–35

Newberg LA, Milde JH, Michenfelder JD (1983) The cerebral metabolic effects of isoflurane at and above concentrations that suppress cortical electrical activity. Anesthesiology 59: 23–28

— — — (1984) Systemic and cerebral effects of isoflurane-induced hypotension in dogs. Anesthesiology 60: 541–546

— — — (1985) Cerebral and systemic effects of hypotension induced by adenosine or ATP in dogs. Anesthesiology 62: 429–436

Newman B, Gelb AW, Lam AM (1986) The effect of isoflurane-induced hypotension on cerebral blood flow and cerebral metabolic rate for oxygen in humans. Anesthesiology 64: 307–310

Ngai SH, Shirasawa R, Cheney DL (1979) Changes in motor activity and acetylcholine turnover induced by lidocaine and cocaine in brain regions of rats. Anesthesiology 51: 230–234

Nicholas JF, Lam AM (1984) Isoflurane-induced hypotension does not cause impairment in pulmonary gas exchange. Can Anesth Soc J 31: 352–358

Nilsson E, Ingvar DH (1965) Cerebral blood flow during neurolept-analgesia in the cat. Acta Anaesth Scand 10: 47–54

Nilsson L (1971) The influence of barbiturate anesthesia upon the energy state and upon acid-base parameters of the brain in arterial hypotension and in asphyxia. Acta Neurol Scand 47: 233–253

Nilsson L, Siesjö BK (1974) Influence of anesthesia on the balance between production and utilization of energy in the brain. J Neurochem 23: 29–36

— — (1975) The effect of phenobarbitone anaesthesia on blood flow and oxygen consumption in the rat brain. Acta Anaesthesiol Scand [Suppl] 57: 18–24

Nishikawa T, Omote K, Namiki A, Takahashi T (1986) The effect of nicardipine on cerebrospinal fluid pressure in humans. Anesth Analg 65: 507–510

Nordström CH, Rhencrone S (1977) Postischaemic cerebral blood flow and oxygen utilization rate in rats anaesthetized with nitrous oxide or phenobarbital. Acta Physiol Scand 101: 230–240

— Siesjö BK (1978) Influence of phenobarbital on changes in the metabolites of the energy reserve of the cerebral cortex following complete ischaemia. Acta Physiol Scand 104: 271–280

— Rhencrona S (1979) Reduction of cerebral blood flow and oxygen consumption with a combination of barbiturate anesthesia and induced hypothermia in the rat. Acta Anaesthesiol Scand 22: 7–12

Novack GD, Bullock JL, Eiseli JH (1978) Fentanyl: cumulative effects and development of short-term tolerance. Neuropharmacology 17: 77–82

Nugent M, Artru AA, Michenfelder JD (1982) Cerebral metabolic, vascular and protective effects of midazolam maleate. Anesthesiology 56: 172–176

Nunn JF (1959) Controlled respiration in neurosurgical anaesthesia. Anaesthesia 14: 413–414

Nussmeier NA, Arlund C, Slogoff S (1986) Neuropsychiatric complications after cardiopulmonary bypass: Cerebral protection by barbiturate. Anesthesiology 64: 165–170

— Fish KJ (1987) Neuropsychologic dysfunction after cardiopulmonary bypass: A comparison of two institutions. Anesthesiology 67: A14

Oguchi K, Arakawa K, Nelson SR, Samson F (1982) The influence of droperidol, diazepam and physiostigmine on ketamine-induced behavior and brain regional glucose utilization in rat. Anesthesiology 57: 353–358

O'Higgins JW (1970) Air embolism during neurosurgery. Br J Anaesth 42: 459–462

Ohm WW, Cullen BF, Amory DW, Kennedy RD (1975) Delayed seizure activity following enflurane anesthesia. Anesthesiology 42: 367–368

Okuda Y, McDowall DG, Ali MM, Lane JR (1976) Changes in CO_2 responsiveness and in autoregulation of the cerebral circulation during and after halothane-induced hypotension. J Neurol Neurosurg Psychiatry 39: 221–230

Ong BY, MacIntyre CJ, Bose D et al (1986) Phenobarbital prevents loss of cerebral blood flow autoregulation after asphyxia in newborn lambs. Anesth Analg 65: S115

Onodera H, Sato G, Kogure K (1987) GABA and benzodiazepine receptors in the gerbil brain after transient ischaemia: Demonstration by quantitative receptor autoradiography. J Cereb Blood Flow Metab 7: 82–88

Oren RE, Rasool NA, Rubinstein EH (1987) Effect of ketamine on cerebral cortical blood flow and metabolism in rabbits. Stroke 18: 441–444

Ori C, Dam M, Pizzolato G, Battistin L, Giron G (1986) Effects of isoflurane anesthesia on local cerebral glucose utilization in the rat. Anesthesiology 65: 152–156

Orlowski JP, Shiesley D, Vidt DG, Barnett GH, Little JR (1988) Labetalol to control blood pressure after cerebrovascular surgery. Crit Care Med 16: 765–768

Ornstein E, Matteo RS, Schwartz AE (1986) The use of esmolol for deliberate hypotension. Anesthesiology 65: A575

— — Winstein JA, Schwartz AE (1987) A randomised controlled trial of esmolol for deliberate hypotension. Anesthesiology 67: A423

Oshita S, Ishikawa T, Tokutsu Y, Takeshita H (1979) Cerebral circulatory and metabolic stimulation with nitrous oxide in the dog. Acta Anaesth Scand 23: 177–181

Owen H, Spence AA (1984) Etomidate (editorial). Br J Anaesth 56: 555–556

Öwall A, Gordon E, Lagerkranser M, Lindquist C, Rudehill A, Sollevi A (1984) Clinical experience with adenosine for controlled hypotension during cerebral aneurysm surgery. Anesth Analg 66: 229–234

— Järnberg P, Brodin L, Sollevi A (1988a) Effects of adenosine-induced hypotension on myocardial hemodynamics and metabolism in fentanyl anesthetized patients with peripheral vascular disease. Anesthesiology 68: 416–421

— Lagerkranser M, Sollevi A (1988) Effects of adenosine-induced hypotension on myocardial hemodynamics and metabolism during cerebral aneurysm surgery. Anesth Analg 67: 228–232

Pearcy WC, Knott JR, Bjurstrom RO (1957) Studies on nitrous oxide, meperidine and levallorphan with unipolar electroencephalography. Anesthesiology 18: 310–315

Pearl RG, Rosenthal MH, Ashton JPA, Murad F (1983a) Aminophylline potentiates sodium nitroprusside-induced hypotension. Anesthesiology 59: A126

— — — (1983b) Pulmonary vasodilator effects of nitroglycerin and sodium nitroprusside in canine oleic acid-induced pulmonary hypertension. Anesthesiology 58: 514–518

Pearl RG, Lawson CP (1986) Hemodynamic effects of PEEP during continuous venous air embolism in the dog. Anesthesiology 64: 724–729

Pelligrino DA, Miletich DJ, Hoffman WE, Albrecht RF (1984) Nitrous oxide markedly increases cerebral cortical metabolic rate and blood flow in the goat. Anesthesiology 60: 405–412

Pena H, Gaines C, Suess D, Crowell RM, Waggener JD, DeGirolami U (1982) Effect of mannitol on experimental focal ischaemia in awake monkeys. Neurosurgery 11: 477–481

Perkins NAK, Bedford RF (1984) Hemodynamic consequences of PEEP in seated neurological patients—implications for paradoxical air embolism. Anesth Analg 63: 429–432

Perkins-Pearson NAK, Marshall WK, Bedford RF (1982) Atrial pressures in the seated position. Implication for paradoxical air embolism. Anesthesiology 57: 493–497

Peterson DO, Drummond JC, Todd MM (1984) Effects of halothane and isoflurane on somatosensory evoked potentials in man. Anesthesiology 61: A344

Pfeifer G, Oehmen S, Limberg NJ, Schultheiss R (1987) Die Wirkung von Isofluran auf den intrakraniellen Druck. Anaesth Intensivther Notfallmed 22: 214–220

Pfitzner J, McLean AG (1987) Venous air embolism and active lung inflation at high and low CVP: A study in "Upright" anesthetized sheep. Anesth Analg 66: 1127–1134

— — Crawshaw KM (1987) Embolized air collection in the superior vena cava of "upright" sheep. Anesth Analg 66: 1135–1140

Phillis JW, Walter GA, O'Regan MH, Stair RE (1987) Increases in cerebral cortical perfusate adenosine and inosine concentrations during hypoxia and ischemia. J Cereb Blood Flow Metab 7: 679–686

Phirman JR, Shapiro HM (1977) Modification of nitrous oxide-induced intracranial hypertension by prior induction of anaesthesia. Anesthesiology 46: 150–151

Piatt JH, Schiff SJ (1984) High dose barbiturate therapy in neurosurgery and intensive care. Neurosurgery 15: 427–444

Pickerodt VWA, McDowall DG, Coroneos NJ, Keaney NP (1972) Effect of althesin on cerebral perfusion, cerebral metabolism and intracranial pressure in the anaesthetized baboon. Br J Anaesth 44: 751–758

Pierce EC, Lambertsen CJ, Deutsch S, Chase PE, Linde HW, Dripps RD, Price HL (1962) Cerebral cinculation and metabolism during thiopental anesthesia and hyperventilation in man. J Clin Invest 41: 1664–1671

Pinaud M, Souron R, Lelausque J, Gazeau MF, Lajat Y, Dixneuf B (1989) Cerebral blood flow and cerebral oxygen consumption during nitroprusside-induced hypotension to less than 50 mm Hg. Anesthesiology 70: 255–260

Pinaud M, Lelausque JN, Fauchoux N, Chetanneau A (1988) Effects of propofol on cerebral hemodynamics and metabolism in patients with head trauma. Anesthesiology 69: A569

Pollay M, Fullenwider C, Roberts A, Stevens FA (1983) Effect of mannitol and furosemide on blood–brain osmotic gradient and intracranial pressure. J Neurosurg 59: 945–950

Poulton TJ, James FM (1979) Cough suppression by lidocaine. Anesthesiology 50: 470–472

Presson RG, Kirk KR, Haselby KA, Wagner WW (1989) The fate of air emboli in the pulmonary circulation of the dog. Anesth Analg 68: S227

Preziosi P, Vacca M (1982) Etomidate and corticotropic axis. Arch Intern Pharmacodyn 256: 308–310

Prior JGL, Hinds CJ, Williams J, Prior PE (1983) The use of etomidate in the management of severe head injury. Intensive Care Med 9: 313–320

Prior PF, Maynard DE, Brierley JB (1978) EEG monitoring for the control of anaesthesia produced by the infusion of althesin in primates. Br J Anaesth 50: 993–1001

Procaccio F, Bingham RM, Hinds CJ, Prior PF (1988) Continuous EEG and ICP monitoring as a guide to the administration of althesin sedation in severe head injury. Intensive Care Med 14: 148–155

Prough DS, Stullken EH, Scuderi PE, Johnson JC (1983a) Intracranial hypertension limits the systemic hypotensive effect of nitroprusside, but not nitroglycerin. Anesthesiology 59: A354

— — — (1983b) Technique of administration alters increases in intracranial hypertension produced by nitroglycerin but not by nitroprusside. Anesthesiology 59: A353

Puchstein C, van Aken H, Anger C, Thys J, Lawin P (1983) Influence of ATP induced hypotension on intracranial pressure and intracranial compliance. Anesthesiology 59: A356

Rao TLK, Mummaneni N, El-Etr AA (1982) Convulsions: An unusual response to intravenous fentanyl administration. Anesth Analg 61: 1020–1021

Rapoport SI, Robinson PJ (1986) Tight-junctional modification as the basis of osmotic opening of the blood–brain barrier. Ann NY Acad Sci 481: 250–266

Rasmussen NJ, Rosendal T, Overgaard J (1978) Althesin in neurosurgical patients: Effects on cerebral haemodynamics and metabolism. Acta Anaesthesiol Scand 22: 257–269

Ratcheson RA, Bilezikjian L, Ferrendelli JA (1977) Effect of nitrous oxide anesthesia upon cerebral energy metabolism. J Neurochem 28: 223–225

Ravussin P, Archer DP, Meyer E, Abou-Madi M, Yamamoto L, Trop D (1985) The effects of rapid infusions of saline and mannitol on cerebral blood volume and intracranial pressure in dogs. Can Anaesth Soc J 32: 506–515

— — Tyler JL, Meyer E et al (1986a) Effects of rapid mannitol infusion on cerebral blood volume. J Neurosurg 64: 104–113

— Chiolero R, Buchser E, DeTribolet N, Freeman J (1986b) CSF pressure changes following mannitol in patients undergoing craniotomy. Anesthesiology 65: A303

— Guinard JP, Ralley F, Thorin D (1988a) Effect of propofol on cerebrospinal fluid pressure and cerebral perfusion pressure in patients undergoing craniotomy. Anaesthesia 43 (Suppl): 37–41

— Berger-Bayer M, Nydegger M, Freeman J (1988b) Thiopentone–isoflurane vs propofol in neuroanesthesia for intracranial surgery. Anesthesiology 69: A577

Ray KF, Kohlenberger RW, Shapiro HM (1979) Local cerebral blood flow and metabolism during halothane and enflurane. Anesthesiology 51: S10

Reicher D, Bhalla P, Rubinstein EH (1987) Cholinergic cerebral vasodilator effect of ketamine in rabbits. Stroke 18: 445–449

Reinhold M, DeRood M, Capon A, Mouawad E, Frèhling J, Verbist A (1974) The Action of enflurane (Ethrane) on cerebral blood flow. Acta Anaesthesiol Belg 25: 257–265

Reinhold H, DeRood M (1976) Cerebral blood flow under enflurane anesthesia. Acta Anaesthesiol Belg 27 [Suppl]: 250–258

Renou AM, Vernhiet J, Orgogozo JM, Caille JM (1976) Effets de l'alfatesine (CT 1341) sur le debit sanguin et le metabolisme cerebral chez l'homme, modifications globales er regionales. Ann Anesthesiol Fr 17: 1247–1254

— — Macrez P et al (1978) Cerebral blood flow and metabolism during etomidate anaesthesia in man. Br J Anaesth 50: 1047–1051

Reulen HJ, Graham R, Spatz M, Klatzo I (1977) Role of pressure gradients and bulk flow in dynamics of vasogenic edema. J Neurosurg 46: 24–35

Richardson HF, Coles BC, Hall GE (1937) Experimental gas embolism: I. Intravenous air embolism. Can Med Ass J 36: 584–588

Roald OK, Forsman M, Steen PA (1988) Partial reversal of the cerebral effects of isoflurane in the dog by benzodiazepine antagonist flumazenil. Acta Anaesthesiol Scand 32: 209–212

— — — (1989) The effect of prolonged isoflurane anaesthesia on cerebral blood flow and metabolism in the dog. Acta Anaesthesiol Scand 33: 210–213

Roberts PA, Pollay M, Engles C, Pendleton B, Reynolds E, Stevens FA (1987) Effect on intracranial pressure of furosemide combined with varying doses and administration rates of mannitol. J Neurosurg 66: 440–446

Rockoff MA, Marshall LF, Shapiro HM (1979) High-dose barbiturate therapy in humans: A clinical review of 60 patients. Ann Neurol 6: 194–199

Rogers MC, Traystman RJ (1979a) Nitroglycerin and nitroprusside induced changes in cerebral hemodynamics. Anesthesiology 51: S199

— Hamburger C, Owen K, Epstein MH (1979b) Intracranial pressure in the cat during nitroglycerin-induced hypotension. Anesthesiology 51: 227–229

Rolly G, Van Aken J (1979) Influence of enflurane on cerebral blood flow in man. Acta Anaesthesiol Scand 71 [Suppl]: 59–63

Ropper AH, Kofke WA, Bromfield EB, Kennedy SK (1986) Comparison of isoflurane, halothane, and nitrous oxide in status epilepticus. Ann Neurology 19: 98–99

Rosa G, Orfie P, Sanfilippo M, Vilardi V, Gasparetto A (1986a) The effects of atracurium besylate (Tracrium) on intracranial pressure and cerebral perfusion pressure. Anesth Analg 65: 381–384

— Sanfilippo M, Viardi V, Orfei P, Gasparetto A (1986b) Effects of vecuronium bromide on intracranial pressure and cerebral perfusion pressure. Br J Anaesth 58: 437–440

Rosner MJ, Coley I (1987) Cerebral perfusion pressure: A hemodynamic mechanism of mannitol and postmannitol hemogram. Neurosurgery 21: 147–156

Roth S, Jones SC, Ebrahim Z, Friel H, Little JR (1986) Cerebral blood flow and metabolism during isoflurane-induced hypotension. Anesthesiology 65: A572

Rudehill A, Lagerkranser M, Lindquist C, Gordon E (1983) Effects of mannitol on blood volume and central hemodynamics in patients undergoing cerebral aneurysm surgery. Anesth Analg 62: 875–880

Rung GW, Hillman DR, Thompson WR, Davis NJ (1987) The effect of bupivacaine scalp infiltration on the hemodynamic response to craniotomy under general anaesthesia. Anesthesiology 67: A432

Safar P, Stezoski W, Nemoto EM (1976) Amelioration of brain damage after 12 minutes cardiac arrest in dogs. Arch Neurol 33: 91–95

Safo Y, Greenberg J, Young M et al (1983) Effects of high dose fentanyl on regional cerebral blood flow. Anesthesiology 59: A306

Safwat AM, Daniel D (1983) Grand mal seizures after fentanyl administration. Letter to editor. Anesthesiology 59: 79

Saintz JJG, Camiruaga JAE, Cano FF, De la Herran JL (1988) Effects of isoflurane on intraventricular pressure in neurosurgical patients. Br J Anaesth 61: 347–349

Sakabe T, Maekawa T, Ishikawa T, Takeshita H (1974) The effects of lidocaine on canine cerebral metabolism and circulation related to the electroencephalogram. Anesthesiology 40: 433–441

— (1975) Effect of enflurane (Ethrane) on canine cerebral metabolism and circulation. Masui 24: 323–331

— Kuramoto T, Inoue S, Takeshita H (1976) Cerebral responses to the addition of nitrous oxide to halothane in man. Br J Anaesth 48: 957–961

— — — Takeshita H (1978) Cerebral effects of nitrous oxide in the dog. Anesthesiology 48: 195–200

— Maekawa T, Fujii S, Ishikawa T, Tateishi A, Takeshita H (1983) Cerebral circulation and metabolism during enflurane anesthesia in humans. Anesthesiology 59: 532–536

— Tsutsui T, Maekawa T, Ishikawa T, Takeshita H (1985) Local cerebral glucose utilization during nitrous oxide and pentobarbital anesthesia in rats. Anesthesiology 63: 262–266

Sale JP (1984) Prevention of air embolism during sitting neurosurgery. The use of inflatable venous neck tourniquet. Anaesthesia 39: 795–799

Samra SK, Deutsch G, Arens JF (1988) Effect of nitrous oxide on global and regional cerebral blood flow in humans. Anesthesiology 69: A536

Sapirstein LA, Ogden E (1956) Theoretic limitations of the nitrous oxide method for the determination of regional blood flow. Circ Res 4: 245–249

Sari A, Okuda Y, Takeshita H (1972) The effects of thalamonal on cerebral circulation and oxygen consumption in man. Br J Anaesth 44: 330–334

— Maekawa T, Tohjo M, Okuda Y, Takeshita H (1976) Effects of althesin on cerebral blood flow and oxygen consumption in man. Br J Anaesth 48: 545–550

Satinover I, Hoffman WE, Albrecht RF, Miletich DJ, Gans BJ (1981) Controlled hypotension with ATP and sodium nitroprusside. Anesthesiology 55: A10

Saul TG, Ducker TB (1982) Effect of intracranial pressure monitoring and aggressive treatment on mortality in severe head injury. J Neurosurg 56: 498–503

Scheller MS, Todd MM, Drummond JC (1984) The effects of halothane and isoflurane on cerebral blood flow at various levels of $PaCO_2$ in rabbits. Anesthesiology 61: A528

— — — (1986a) Isoflurane, halothane, and regional cerebral blood flow at various levels of $PaCO_2$ in rabbits. Anesthesiology 64: 598–604

— Drummond JC, Todd MM, Shapiro H, Zornow MH (1986b) Are recommendations regarding barbiturate protection during bypass justified? Anesthesiology 65: 230–231

— Todd MM, Drummond JC, Zornow MH (1987) The intracranial pressure effects of isoflurane and halothane administered following cryogenic brain injury in rabbits. Anesthesiology 67: 507–512

— Tateichi A, Drummond JC, Zornow MH (1988) The effects of sevoflurane on cerebral blood flow, cerebral metabolic rate for oxygen, intracranial pressure and electroencephalogram are similar to those of isoflurane in the rabbit. Anesthesiology 68: 548–551

Schettini A, Moreshead G (1978) Effects of halothane and sodium thiopentone on surface brain pressure and brain electrical impedance in dogs with normal intracranial tension. Br J Anaesth 50: 1003–1012

— Furniss WW (1979) Brain water and electrolyte distribution during the inhalation of halothane. Br J Anaesth 51: 1117–1123

— Stahurski B, Young HF (1982) Osmotic and osmotic-loop diuresis in brain surgery. Effects on plasma and CSF electrolytes and ion excretion. J Neurosurg 56: 679–684

Schubert A, Pterson DO, Drummond JC; Saidman LJ (1986) The effect of high-dose fentanyl on human median nerve somatosensory evoked responses. Anesth Analg 65: S136

Schubert J, Brill WA (1968) Antagonism of experimental cyanide toxicity in relation to the *in vivo* activity of cytochrome oxidase. J Pharmacol Exp Ther 162: 352–359

Schulte am Esch J, Pfeifer G, Thiemig I (1978) Effects of etomidate and thiopentone on the primarily elevated intracranial pressure (ICP). Anaesthesist 27: 71–75

Schulte am Esch J, Pfeiffer G, Hiemig I, Entzian W (1978) The influence of intravenous anaesthetic agents on primarily increased intracranial pressure. Acta Neurochir (Wien) 45: 15–25

— Thiemig I, Pfeifer G, Entzian W (1979) Die Wirkung einiger Inhalationsanaesthetika auf den Intrakraniellen Druck unter besonderer Berücksichtigung des Stickoxydul. Anaesthesist 28: 136–141

Schultz KD, Schultz K, Schultz G (1977) Sodium nitroprusside and other smooth muscle relaxants increase cyclic GMP levels in art ductus deferens. Nature 256: 750–751

Schulz V, Gross R, Pasch T, Busse J, Loeschcke G (1982) Cyanide toxicity of sodium nitroprusside in therapeutic use with and without sodium thiosulphate. Klin Wochenschr 60: 1393–1400

Schwedler M, Miletich DJ, Albrecht RF (1982) Cerebral blood flow and metabolism following ketamine administration. Can Anaesth Soc J 29: 222–226

Sear JW, Walters FJM, Wilkins DG, Willatts SM (1984) Etomidate by infusion for neuroanaesthesia. Kinetic and dynamic interactions with nitrous oxide. Anaesthesia 39: 12–18

Sebel AM, Bovill JG (1983) Fentanyl and convulsions. Anesth Analg 62: 858–859

Sebel PS, Bovill JG, Wauquier A, Rog P (1981) Effects of high-dose fentanyl anesthesia on the electroencephalogram. Anesthesiology 55: 203–211

— Ingram DA, Flynn PJ, Rutherfoord CF, Rogers H (1986) Evoked potentials during isoflurane anaesthesia. Br J Anaesth 58: 580–585

Seki H, Ogawa A, Yoshimoto T, Suzuki J (1981) Effect of mannitol on rCBF in canine thalamic ischemia. An experimental study. Brain Nerve (Tokyo) 33: 1101–1105

Selman RW, Spetzler RF, Roessmann UR, Rosenblatt JI, Crumrine RC (1981)

Barbiturate infusion coma therapy for focal cerebral ischaemia. J Neurosurg 55: 220–226

Seo K, Maekawa T, Takeshita H, Okuda Y (1984) Cerebral energy state and glycolytic metabolism during enflurane anesthesia in the rat. Acta Anaesthesiol Scand 28: 215–219

Seyde WC, Longnecker DE (1986) Cerebral oxygen tension in rats during deliberate hypotension with sodium nitroprusside, 2-chloroadenosine or deep isoflurane anesthesia. Anesthesiology 64: 480–485

— Ellis JE, Longnecker DE (1986) The addition of nitrous oxide to halothane decreases renal and splanchnic flow and increases cerebral blood flow in rats. Br J Anaesth 58: 63

Shafer A, White PF, Schütler J, Rosenthal MH (1983) Use of a fentanyl infusion in the intensive care unit: Tolerance to its anesthetic effects. Anesthesiology 59: 245–248

Shah NK, Long CW, Marx W, Diresta GR, Arbit E, Mascott C, Mallya K, Bedford RF (1988) Cerebrovascular response to CO_2 in edematous brain during either fentanyl or isoflurane anesthesia. Anesthesiology 69: A620

Shapiro HM, Wyte SR, Harris AB, Galindo A (1972a) Acute intraoperative intracranial hypertension in neurosurgical patients. Anesthesiology 37: 399–405

— — — (1972b) Ketamine anaesthesia in patients with intracranial pathology. Br J Anaesth 44: 1200–1204

— Galindo A, Wyte SR, Harris AB (1973) Rapid intraoperative reduction of intracranial pressure with thiopental. Br J Anaesth 45: 1057–1061

— Wyte SR, Loeser J (1974) Barbiturate augmented hypothermia for reduction of persistent intracranial hypertension. J Neurosurg 40: 90–100

— Greenberg JH, Reivich M, Ashmead G, Sokoloff L (1978) Local cerebral glucose uptake in awake and halothane-anesthetized primates. Anesthesiology 48: 97–103

— (1985) Barbiturates in brain ischaemia. Br J Anaesth 57: 82–95

— (1986) Anaesthesia effects upon cerebral blood flow, cerebral metabolism, electroencephalogram, and evoked potentials. In: Miller DR (ed) Anesthesia. Churchill Livingstone, New York Edinburgh London, 2: pp 1249–1288

Sharbrough FW, Messick JM, Sundt TM Jr (1973) Correlation of continuous electroencephalograms with cerebral blood flow measurements during carotid endarterectomy. Stroke 4: 674–683

Shenkin HA, Goluboff B, Haft H (1962) The use of mannitol for the reduction of intracranial pressure in intracranial surgery. J Neurosurg 19: 897–901

Shibutani K, Komatsu T, Bhalodia R, Kubal C, Taddonio R (1983) Rebound hemodynamic responses after the withdrawal of nitroprusside in anesthetized patients. Anesthesiology 59: A16

Shimosato S, Carter JG, Kemmotsu O, Takahashi T (1982) Cardo-circulatory effects of prolonged administration of isoflurane in normocarbic human volunteers. Acta Anaesthesiol Scand 26: 27–30

Shokunbi MT, Gelb AW, Peerless SJ, Mervart M, Floyd P (1986) An evaluation of the effect of lidocaine in experimental focal cerebral ischemia. Stroke 17: 962–966

Shokunbi MT, Gelb AW, Miller DJ, Wu XM (1987) A continuous infusion of lidocaine protects in temporary focal cerebral ischaemia. Anesthesiology 67: A580

Shupak RC, Harp JR, Stevenson-Smith W et al (1983) High-dose fentanyl for neuroanesthesia. Anesthesiology 58: 579–582

— — (1985) Comparison between high-dose sufentanil-oxygen and high-dose fentanyl-oxygen for neuroanesthesia. Br J Anaesth 57: 375–381

Sidi A, Cotev S, Hadani M, Wald U, Feinsod M, Perel A (1983) Long-term barbiturate infusion to reduce intracranial pressure. Crit Care Med 11: 478–481

Siesjö BK (1984) Cerebral circulation and metabolism. J Neurosurg 60: 883–908

Sinclair ME, Sear JW, Summerfield RJ, Fisher A (1988) Alfentanil infusions in the intensive therapy unit. Intensive Care Med 14: 55–59

Sink JD, Comer BP, James PM, Loveland SR (1976) Evaluation of catheter placement in the treatment of venous air embolism. Ann Surg 183: 58–61

Sklar FH, Beyer CW, Ramanathan M, Clark WK (1980) The effects of furosemide on CSF dynamics in patients with pseudotumor cerebri. In: Shulman K, Marmarou A, Miller JD, Becker DP, Hochwald GM, Brock M (eds) Intracranial pressure IV. Springer, Berlin Heidelberg New York, pp 660–663

Smith AL, Neigh JL, Hoffman JC, Wollman H (1970) Effects of general anesthesia on autoregulation of cerebral blood flow in man. J Appl Physiol 29: 665–669

— Wollman H (1972) Cerebral blood flow and metabolism. Effects of anesthetic drugs and techniques. Anesthesiology 36: 378–400

— (1973) The mechanism of cerebral vasodilatation by halothane. Anesthesiology 39: 581–587

— Hoff JT, Nielsen SL, Larson CP (1974) Barbiturate protection in acute focal cerebral ischemia. Stroke 5: 1–7

— Marque JJ (1976) Anesthetics and cerebral oedema. Anesthesiology 45: 64–72

Smith DS, Rehncrona S, Siesjö BK (1980) Inhibitory effects of different barbiturates on lipid peroxidation in the brain tissue in vitro: Comparison with the effects of promethazine and chlorpromazine. Anesthesiology 53: 186–194

Smith NT, Bickford RG, Sanford TJ, Dec-Silver H, Blasco TA (1987) Do seizures occur during clinical induction with opiates? Anesthesiology 67: A385

Snyder BD, Ramirez-Lassepas M, Sukhum P, Fryd D, Sung JH (1979) Failure of thiopental to modify global anoxic injury. Stroke 10: 135–141

Sollevi A, Lagerkranser M, Irestedt L, Gordon E, Lindquist C (1984a) Controlled hypotension with adenosine in cerebral aneurysm surgery. Anesthesiology 61: 400–405

— — Andreen M, Irestedt L (1984b) Relationship between arterial and venous adenosine levels and vasodilatation during ATP- and adenosine-infusion in dogs. Acta Physiol Scand 120: 171–176

— Ericson K, Eriksson L, Lindquist S, Lagerkranser M, Stone-Elander S (1987) Effect of adenosine on human cerebral blood flow as determined by positron emission tomography. J Cereb Blood Flow Metab 7: 673–678

Sokoll MD, Kassell NF, Davles LR (1982) Large dose thiopental anesthesia for intracranial aneurysm surgery. Neurosurgery 10: 555–562

Spargo PM, Tait AR, Knight PR, Kling TF (1987) Effect of nitroglycerin-induced hypotension on canine spinal cord blood flow. Br J Anaesth 59: 640–647

Sperry RJ, Monk CR, Durieux ME et al (1987) Regional cerebral blood flow during deliberate hypotension complicated by hemorrhage. Anesthesiology 67: A575

Spetzler RF, Wilson CB, Weinstein P et al (1978) Normal perfusion pressure breakthrough theory. In: Clinical neurosurgery, vol 26. Williams and Wilkins, pp 651–672

— Selman WR, Roski RA, Bonstelle C (1982) Cerebral revascularization during barbiturate coma in primates and humans. Surg Neurol 17: 111–115

Spiess BD, Braverman B, Woronowicz A, Ivankovich AD (1986) Protection from cerebral air emboli with perfluorocarbons in rabbits. Anesthesiology 65: A361

Spiss CK, Zadrobilek E, Weindlmayr-Goettel M, Marian F, Draxler VH (1985) Nifedipine induced hypotension in man: hemodynamic response during isoflurane and halothane anesthesia. Anesthesiology 63: A93

Standefer M, Bay JW, Trusso R (1984) The sitting position in neurosurgery: A retrospective analysis of 488 cases. Neurosurgery 14: 649–658

Standefer M, Little JR (1986) Improved neurological outcome in experimental focal cerebral ischaemia treated with propranolol. Neurosurgery 18: 136–140

Steen PA, Milde JH, Michenfelder JD (1978) Cerebral metabolic and vascular effects of barbiturate therapy following complete global ischaemia. J Neurochem 31: 1317–1324

— — — (1979) No barbiturate protection in a dog model of complete cerebral ischemia. Ann Neurol 5: 343–349

— Michenfelder JD (1979) Barbiturate protection in tolerant and nontolerant hypoxic mice. Anesthesiology 50: 404–408

— — (1979) Neurotoxicity of anaesthetics. Anesthesiology 50: 437–453

— Newberg L, Milde JH, Michenfelder JD (1983) Hypothermia and barbiturates: Individual and combined effects on canine cerebral oxygen consumption. Anesthesiology 58: 527–532

Steinbach JJ, Mattar AG, Mahin DT (1976) Alteration of the cerebral blood flow study due to reflux in internal jugular veins. J Nucl Med 17: 61–64

Stephan H, Sonntag H, Schenk HD, Kohlhausen S (1987) Einfluss von Disoprivan (Propofol) auf die Durchblutung und Sauerstoffverbrauch des Gehirns und die CO_2 Reaktivität der Hirngefässe beim Menschen. Anaesthetist 36: 60–65

— — Seyde WC, Henze T, Textor J (1988) Energie-und Aminosären-Stoffwechsel des menschlichen Gehirns unter Desoprivan und verschiedenen $PaCO_2$—Werten. Anaesthesist 37: 297–304

Stevens WC, Cromwell TH, Halsey MJ, Eger EI, Shakespeare TF, Bahlman SH (1971) The cardiovascular effects of a new inhalation anesthetic, forene in human volunteers at constant arterial carbon dioxide tension. Anesthesiology 35: 8–16

Stirt JA, Grosslight KR, Bedford RF, Vollmer D (1986) Pretreatment with a defascilulating dose of metocurine prevents succinylcholine-induced increases in intracranial pressure. Anesthesiology 65: A302

Stockard JJ, Schauble JF, Bickford RG (1972) On-line computer analyses of the electroencephalogram during anesthesia. In: Scientific abstracts of the 1972 Meeting. Am Ass Anaesth 193–194

Stockard JJ, Bickford R (1975) The neurophysiology of anesthesia. In: Gordon E, (ed) A basis and practice of neuroanesthesia. Excerpta medica, Amsterdam, pp 3–46

Stone JG, Khambatta HJ, Matteo RS (1976) Pulmonary shunting during anesthesia with deliberate hypotension. Anesthesiology 45: 508–515

Stoyka WW, Schutz H (1975) The cerebral response to sodium nitroprusside and trimethaphan controlled hypotension. Can Anaesth Soc J 22: 275–283

Strandell T, Lundbergh P (1973) Measurements of 133-Xe activity in blood sampled in plastic and glass syringes. J Clin Lab Invest 32: 89–92

Strandgaard S, Jones JV, MacKenzie ET, Harper AM (1975) Upper-limit of cerebral blood flow autoregulation in experimental renovascular hypertension in the baboon. Circ Res 37: 164–167

— (1976) Autoregulation of cerebral blood flow in hypertensive patients. The modifying influence of profound antihypertensive treatment on the tolerance to acute, drug-induced hypotension. Circulation 53: 720–727

— Paulson OB (1984) Cerebral autoregulation. Stroke 15: 413–416

Stullken EH, Sokoll MD (1975) Anesthesia and subarachnoid intracranial pressure. Anesth Analg 54: 494–500

— — (1975) Intracranial pressure during hypotension and subsequent vasopressor therapy in anesthetized cats. Anesthesiology 42: 425–431

— Milde JH, Michenfelder JD, Tinker JH (1977) The nonlinear responses of cerebral metabolism to low concentrations of halothane, enflurane, isoflurane and thiopental. Anesthesiology 46: 28–34

Stånge K, Lagerkranser M, Rudehill A, Sollevi A (1989) Effect of adenosine-induced hypotension on cerebral blood flow and metabolism in the pig. Acta Anaesthesiol Scand 33: 199–203

Sutherland G, Lesiuk H, Bose R, Sima AAF (1988) Effect of mannitol, nimodopine, and indomethacin singly or in combination on cerebral ischemia in rats. Stroke 19: 571–578

Symon L, Branston NM, Chikovani O (1979) Ischemic brain edema following middle cerebral artery occlusion in baboons: Relationship between regional cerebral water content and blood flow at 1 and 2 hours. Stroke 10: 184–191

Søndergård W (1961) Intracranial pressure during general anaesthesia. Dan Med Bull 8: 18–26

Tagawa H, Vander AJ (1970) Effects of adenosine compounds on renal function and renin secretion in dogs. Circ Res 26: 327–338

Takahashi T, Takasaki M, Namiki A, Dohi S (1973) Effects of althesin on cerebrospinal fluid pressure. Br J Anaesth 45: 179–184

Takasaki M (1974) The effects of enflurane on canine cerebral oxygen consumption and blood flow. Jap J Anesthesiol 23: 806

Takeshita H, Okuda Y, Sari A (1972) The effects of ketamine on cerebral circulation and metabolism in man. Anesthesiology 36: 69–75

— Michenfelder JD, Theye RA (1972) The effects of morphine and N-allyl-normorphine on canine cerebral metabolism and circulation. Anesthesiology 37: 605–612

Tam S, Chung F, Campbell M (1987) Intravenous lidocaine: Optimal time of injection before tracheal intubation. Anesth Analg 66: 1036–1038

Tamasy V, Koranyi L, Tekeres M (1975) EEG and multiple unit activity during ketamine and barbiturate anaesthesia. Br J Anaesth 47: 1247–1251

Tambuniello G, Munari C, Gianesi GC (1978) The behaviour of intracranial pressure during anesthesia with enflurane and oxygen alone. Minerva Anesthesiol 44: 923

Tanaka A, Tomonaga M (1987) Effect of mannitol on cerebral blood flow and microcirculation during experimental middle cerebral artery occlusion. Surg Neurol (28) 189–195

Tanaka GY (1974) Hypertensive reaction to nalaxone. JAMA 228: 25–26

Tans JTJ, Poortvliet DCJ (1983) Intracranial volume-pressure relationship in man. Part 2: Clinical significance of the pressure volume index. J Neurosurg 59: 810–816

Tarkkanen L, Laitinen L, Johansson G (1974) Effects of d-tubocuranine on intracranial pressure and thalamic electrical impedance. Anesthesiology 40: 247–251

Tateichi A, Maekawa T, Takeshita H, Wakuta K (1981) Diazepam and intracranial pressure. Anesthesiology 54: 335–337

Tateishi A, Sano T, Takeshita H, Suzuki T, Tokuno H (1988) Effects of nifedipine on intracranial pressure in neurosurgical patients with arterial hypertension. J Neurosurg 69: 213–215

Tateichi A, Zornow MH, Scheller MS, Canfell PC (1989) Electroencephalographic effects of laudanosine in an animal model of epilepsy. Br J Anaesth 62: 548–552

Tausk HC, Miller R (1983) Anaesthesia for posterior fossa surgery in the sitting position. Bull New York Academy Med 59: 771–783

Theye RA, Michenfelder JD (1968a) The effect of halothane on canine cerebral metabolism. Anesthesiology 29: 1113–1118

— — (1968b) The effect of nitrous oxide on canine cerebral metabolism. Anesthesiology 29: 1119–1124

Thiagarajah S, Azar I, Lear E, Suazo J (1983) Comparative evaluation of intracranial pressure changes in cats during adenosine triphosphate and nitroprusside induced hypotension. Anesthesiology 59: A352

— — — Rudolf D (1986) Effect of diltiazem-induced hypotension on normal and increased intracranial pressure of cats. Can Anaesth Soc J 33: 578–582

Thiagarajah S, Sophie S, Azar I, Lear E (1988) Effect of succinylcholine on the ICP of cats with and without thiopentone pretreatment. Br J Anaesth 60: 157–160

Thilmann J, Zeumer H (1974) Untersuchungen zur Behandlung des Hirnödems mit hohen Dosen Furosemid. Dtsch med Wschr 99: 932–935

Thomas WA, Cole PV, Etherington NJ, Prior PF, Stefansson SB (1985) Electrical activity of the cerebral cortex during induced hypotension in man. Br J Anaesth 57: 134–141

Thompson GE (1972) Ketamine-induced convulsions. Anesthesiology 37: 662–663

Thornton C, Catley DM, Jordan C, Lehane JR, Royston D, Jones JG (1983)

Enflurane anaesthesia causes graded changes in the brainstem and early cortical auditory evoked response in man. Br J Anaesth 55: 479–485

Tindall GT, Craddock A, Greenfield JC (1967) Effects of the sitting position on blood flow in the internal carotid artery of man during general anesthesia. J Neurosurg 26: 383–389

— Greenfield JC Jr (1973) The effects of intra-arterial histamine on blood flow in the internal and external carotid artery of man. Stroke 4: 46–49

Tinker JH, Vandam LD (1972) How effective is the G suit in neurosurgical operations? Anesthesiology 36: 609–611

— Michenfelder JD (1976) Sodium nitroprusside: Pharmacology, toxicology and therapeutics. Anesthesiology 45: 340–353

— Sharbrough FW, Michenfelder JD (1977) Anterior shift of the dominant EEG rhythm during anesthesia in the Java monkey: Correlation with anesthetic potency. Anesthesiology 46: 252–259

Todd MM, Morris PJ, Philbin DM (1980) Acute neurologic changes caused by nitroprusside-induced intracranial hypertension. Anesth Analg 59: 561–562

— Chadwick HS, Shapiro HM, Dunlop BJ, Marshall LF, Dueck R (1982) The neurologic effects of thiopental therapy following experimental cardiac arrest in cats. Anesthesiology 57: 76–86

— Drummond JC, Sang UH (1984a) The hemodynamic consequences of high-dose methohexital anesthesia in humans. Anesthesiology 61: 495–501

— — (1984b) A comparison of the cerebrovascular and metabolic effects of halothane and isoflurane in the cat. Anesthesiology 60: 276–282

— — Sang UH (1985) The hemodynamic consequences of high-dose thiopental anesthesia. Anesth Analg 64: 681–687

— (1987) The effects of PaCO$_2$ on the cerebrovascular response to nitrous oxide in the halothane-anesthetized rabbit. Anesth Analg 66: 1090–1095

— Drummond JC, Sang UH (1987) Hemodynamic effects of high dose pentobarbital: Studies in elective neurosurgical patients. Neurosurgery 20: 559–563

Tommasino C, Maekawa T, Shapiro HM (1984) Fentanyl-induced seizures activate subcortical brain metabolism. Anesthesiology 60: 283–290

— — — (1986) Local cerebral blood flow during lidocaine-induced seizures in rats. Anesthesiology 64: 771–777

Toung TJK, Miyabe M, McShane AJ, Rogers MC, Traystman RJ (1988) Effect of PEEP and jugular venous compression on canine cerebral blood flow and oxygen consumption in the head elevated position. Anesthesiology 68: 53–58

Traeger SM, Henning RJ, Doblin W, Giannotta S, Weil H, Weiss M (1983) Hemodynamic effects of pentobarbital therapy for intracranial hypertension. Crit Care Med 11: 697–701

Trøjaborg W, Boysen G (1973) Relation between EEG, regional cerebral blood flow and internal carotid artery pressure during carotid endarterectomy. EEG Clin Neurophysiol 34: 61–69

Tuman KJ, Spiess BD, McCarthy RJ, Ivankovich AD (1986) Cardiorespiratory effects of venous air embolism in dogs receiving a perfluorocarbon emulsion. J Neurosurg 65: 238–244

Turner JM, Coroneos NJ, Gibson RM, Powell D, Ness MA, McDowall DG (1973) The effect of althesin on intracranial pressure in man. Br J Anaesth 45: 168–171

— Powell D, Gibson RM, McDowall DG (1977) Intracranial pressure changes in neurosurgical patients during hypotension induced with sodium nitroprusside or trimethaphan. Br J Anaesth 49: 419–425

TySmith N, Quinn ML, Westover JC et al (1986) The EEGs of alfentanil, sufentanil, and fentanyl: A quantitative comparison. Anesth Analg 65: S145

Unni VKN, Young HSA (1986) Effects of atracurium on intracranial pressure in man. Anaesthesia 41: 1047–1049

Van Aken J, Rolly G (1976) Influence of etomidate, a new short acting anaesthetic agent, on cerebral blood flow in man. Acta Anaesthesiol Belg [Suppl] 27: 175–180

— Hautekiet A, Rolly C (1977) Influence of enflurane on cerebral blood flow in man. Acta Anaesthesiol Belg 28: 133–140

— Puchstein C, Fitch W, Graham DI (1984a) Haemodynamic and cerebral effects of ATP-induced hypotension. Br J Anaesth 56: 1409–1415

— — Anger C, Heinecke A, Lawin P (1984b) Changes in intracranial pressure and compliance during adenosine triphosphate-induced hypotension in dogs. Anesth Analg 63: 381–385

— Fitch W, Graham DI, Brussel T, Themann H (1986) Cardiovascular and cerebrovascular effects of isoflurane-induced hypotension in the baboon. Anesth Analg 65: 565–574

Vandesteene A, Trempont V, Engelman E, Deloof T, Focroul M, Schoutens A, DeRood M (1988) Effect of propofol on cerebral blood flow and metabolism in man. Anaesthesia 43 [Suppl]: 42–43

Van Hemelrijck J, Van Aken H, Plets C, Goffin J (1988) The effects of propofol on ICP and cerebral perfusion pressure in patients with brain tumors. Anesthesiology 69: A570

Van Reempts J, Borgers M, Van Eyndhoven J, Hermans C (1982) Protective effects of etomidate in hypoxic-ischemic brain damage in the rat. A morphologic assessment. Experimental Neurology 76: 181–195

Vernhiet J, Renou AM, Orgogozo JM, Constant P, Caille JM (1978a) Effects of diazepam-fentanyl mixture on cerebral blood flow and oxygen consumption in man. Br J Anaesth 50: 165–169

Vesely R, Hoffman WE, Gil KSL (1986) Cerebrovascular effects of curare and histamine in the rat. Anesthesiology 65: A336

Vesey CJ, Cole PV, Simpson PJ (1976) Cyanide and thiocyanate concentrations following sodium nitroprusside infusion in man. Br J Anaesth 48: 651–660

Virtue RW, Lund LO, Phleps M, Vogel JHK, Beckwitt H, Heron M (1966) Difluoromethyl 1,1,2-trifluoro-2-chloroethyl ether as an anesthetic agent: results with dogs and a preliminary note on observations in man. Can Anaesth Soc J 13: 233–241

Voorhies RM, Fraser RAR, Poznak AV (1983) Prevention of air embolism with positive end expiratory pressure. Neurosurgery 12: 503–506

Waaben J, Husum B, Hansen AJ, Gjedde A (1989) Hypocapnia prevents the decrease in regional cerebral metabolism during isoflurane-induced hypotension. J Neurosurg Anesthesiol 1: 29–34

— — — — (1989) Regional cerebral blood flow and glucose utilization during hypocapnia and adenosine-induced hypotension in the rat. Anesthesiology 70: 299–304

Wagman IH, DeJong RH, Prince DA (1967) Effects of lidocaine on the central nervous system. Anesthesiology 28: 155–172

Wagner RL, White PF, Kan PB, Rosenthal MH, Feldman D (1984) Inhibition of adrenal steroidogenesis by the anesthetic etomidate. N Engl J Med 310: 1415–1421

— — (1984) Etomidate inhibits adrenocortical function in surgical patients. Anesthesiology 61: 647–651

Wahl M, Kuschinsky W (1976) The dilatory action of adenosine on pial arteries of cats and its inhibition by theophylline. Pflügers Arch 362: 55–59

Wallin RF, Regan BM, Napoli MD, Stern IJ (1975) Sevoflurane: a new inhalation anaesthetic agent. Anesth Analg 54: 758–766

Ward JD, Becker DP, Miller JD, Choi SC, Marmarou A, Wood C, Newlon PG, Keenan R (1985) Failure of prophylactic barbiturate coma in the treatment of severe head injury. J Neurosurg 62: 383–388

Ward RJ, Danziger F, Bonica JJ, Allen GD, Tolas AG (1966) Cardiovascular effects of change in posture. Aerospace Medicine 37: 257–259

Warner DS, Boarini DJ, Kassell NF (1985) Cerebrovascular adaptation to prolonged halothane anesthesia is not related to cerebrospinal fluid pH. Anesthesiology 63: 243–248

— Deshpanda JK, Wieloch T (1986) The effect of isoflurane on neuronal necrosis following near-complete forebrain ischaemia in the rat. Anesthesiology 64: 19–23

— Godersky JC, Smith M-L (1988) Failure of pre-ischaemic lidocaine administration to ameliorate global ischaemic brain damage in the rat. Anesthesiology 68: 73–78

Watanaba T, Yoshimoto T, Ogawa A, Sakamoto T, Suzuki J (1979) The effect of mannitol in preserving the development of cerebral infarction. An electron microscopic investigation. Neurol Surg (Tokyo) 7: 859–866

Wauquier A, Van den Broeck WAE, Verheyen JL, Janssen PAJ (1978) Electroencephalographic study of the short-acting hypnotics etomidate and methohexital in dogs. Europ J Pharmacol 47: 367–377

— Achton D, Clincke G, Niemegers CJE (1981) Anti-hypoxic effects of etomidate, thiopental and methohexital. Arch Int Pharmacodyn Ther 249: 330–334

— (1982) Brain protective properties of etomidate and flunarizine. J Cereb Blood Flow Metab 2 [Suppl 1]: S53–S56

Wechsler RL, Dripps RD, Kety SS (1951) Blood flow and oxygen consumption of the human brain during anesthesia produced by thiopental. Anesthesiology 12: 308–313

Weiss MH, Wertman N, Apuzzo MLJ, Heiden JS, Kurze T (1977) The influence of myoneural blockers in intracranial dynamics. Bull Los Angel Neurologic Soc 42: 1–7

Weitzner SW, McCay GT, McCay T, Binder LS (1963) Effects of morphine, levallorphan, and respiratory gases on increased intracranial pressure. Anesthesiology 24: 291–298

White PF, Schlobohm RM, Pitts LH, Lindauer JM (1982) A randomized study of drugs for preventing increases in intracranial pressure during endotracheal suctioning. Anesthesiology 57: 242–244

Wiklund L (1986) Reversal of sedation and respiratory depression after anaesthesia by the combined use of physiostigmine and nalaxone in neurosurgical patients. Acta Anaesthesiol Scand 30: 374–378

Wildsmith JAW, Drummond GB, MacRae WR (1979) Metabolic effects of induced hypotension with trimethaphan and sodium nitroprusside. Br J Anaesth 51: 875–879

Wilkinson IMS, Browne DRG (1970) The influence of anaesthesia and of arterial hypocapnia on regional blood flow in the normal human cerebral hemisphere. Br J Anaesth 42: 472–482

Winn HR, Rubio R, Berne RM (1979) Brain adenosine production in the rat during 60 seconds of ischemia. Circ Res 45: 486–492

— Welsh JE, Rubio R, Berne RM (1980a) Changes in brain adenosine during bicuculline-induced seizures in rats. Circ Res 47: 568–577

— — — — (1980b) Brain adenosine production in the rat during sustained alteration in systemic blood pressure. Am J Physiol 239: H636–641

— Rubio R, Berne RM (1981) The role of adenosine in the regulation of cerebral blood flow. J Cereb Blood Flow Metab 1: 239–244

Winters WD (1972) Epilepsia or anaesthesia with ketamine. Anesthesiology 36: 309–312

Wise BL, Chater N (1962) The value of hypertonic mannitol solution in decreasing brain mass and lowering cerebrospinal-fluid pressure. J Neurosurg 19: 1038–1043

Wolffe JB, Robertson HF (1935) Experimental air embolism. Ann Intern Med 9: 162–165

Wollman H, Alexander SC, Cohen PJ, Chase PE, Melman E, Behar MG (1964) Cerebral circulation of man during halothane anesthesia. Effects of hypocarbia and of d-tubocurarine. Anesthesiology 25: 180–184

— — — Smith TC, Chase PE, Van der Molen RA (1965) Cerebral circulation during general anesthesia and hyperventilation in man. Thiopental induction to nitrous oxide and d-tubocuranine. Anesthesiology 26: 329–334

— Smith AL, Neigh JL, Hoffman JC (1969) Cerebral blood flow and oxygen consumption in man during electroencephalographic seizure patterns associated with ethrane anesthesia. In: Brock M, Fieschi C, Ingvar D et al (eds) Cerebral blood flow—clinical and experimental results. Springer, New York, pp 246–248

Wollman SB, Orkim LR (1968) Postoperative human reaction time and hypocapnia during anesthesia. Br J Anaesth 40: 920–927

Wood EH, Lambert EH (1952) Some factors which influence protection afforded by pneumatic anti-G suits. J Aviation Med 23: 218–228

Woodcock TE, Murkin JM, Farrar K, Tweed A, Guiraudon GM, McKenzie N (1987) Pharmacologic EEG suppression during cardiopulmonary bypass: Cerebral hemodynamic and metabolic effects of thiopental or isoflurane during hypothermia and normothermia. Anesthesiology 67: 218–224

Woodside JR, Garner L, Bedford RF, Miller ED, Longnecker DE, Epstein RM (1982) Captopril reduces the dose requirement for SNP-induced hypotension. Anesthesiology 57: A61

Wu PH, Phillis JW (1982) Uptake of adenosine by isolated rat brain capillaries. J Neurochem 38: 687–690

Wyrwicz AM, Pszenny MH, Schofield JC, Tillman PC, Gordon RE, Marin PA (1983) Noninvasive observations of fluorinated anesthetics in rabbit brain by fluorine-19 nuclear magnetic resonance. Science 222: 428–430

Wyte SR, Shapiro HM, Turner P, Harris AB (1972) Ketamine-induced intracranial hypertension. Anesthesiology 36: 174–176

Yahagi N, Furuya H (1987) The effects of halothane and pentobarbital on the threshold of transpulmonary passage of venous air emboli in dogs. Anesthesiology 67: 905–909

Yano M, Ikeda Y, Kobayashi S, Yamamoto Y, Otsuka T (1986) The outcome with barbiturate therapy in severe head injuries. In: Miller JD, Teasdale GM, Rowan JO, Galvraith SL, Mendelow AD (eds) Intracranial pressure VI. Springer, Berlin Heidelberg New York, pp 769–773

Yano M, Nishiyama H, Yokota H et al (1986) Effect of lidocaine on ICP response to endotracheal suctioning. Anesthesiology 64: 651–653

Yate PM, Thomas D, Sebel P (1984) Alfentanil infusion for sedation and analgesia in intensive care. Lancet ii: 396–397

Yatsu FM, Diamond I, Graziano C, Linquist P (1972) Experimental brain ischemia: Protection from irreversible damage with a rapid-acting barbiturate (methohexital). Stroke 3: 726–732

Yonas H, Dujovny M, Nelson D, Lipton SD, Segel R, Agdeppa D, Mazel M (1981) The controlled delivery of thiopental and delayed cerebral revascularization. Surg Neurol 15: 27–34

Young ML, Smith DS, Greenberg J, Reivich M, Harp JR (1984) Effects of sufentanil on regional cerebral glucose utilization in rats. Anesthesiology 61: 564–568

— — Murtagh F, Vasquez F, Levitt J (1986) Comparison of surgical and anesthetic complications in neurosurgical patients experiencing venous air embolism in the sitting position. Neurosurgery 18: 157–161

Young W, Flamm ES, Demopoulus HB, Tomasula JJ, DeCrescito V (1981) Effect of nalaxone on posttraumatic ischaemia in experimental spinal contusion. J Neurosurg 55: 209–219

Zasslow MA, Pearl RPG, Larson CP, Silverberg G (1986) PEEP does not affect left atrial-right atrial pressure gradient in neurosurgical patients. Anesthesiology 65: A304

Zattoni J, Siani C, Rivano C (1974) Effects of ethrane on intracranial pressure. In: Lawin P, Beer R (eds) Ethrane: Proceedings of the first European Symp on modern Anaesthetic Agents. Springer, Berlin Heidelberg New York, 272 pp

Zornow MH, Todd MM, Moore SS (1987) The acute cerebral effects of changes in plasma osmolality and oncotic pressure. Anesthesiology 67: 936–941

Zubrow AB, Daniel SS, Stark RI, Husain MK, James LS (1983) Plasma renin, catecholamine, and vasopressin during nitroprusside-induced hypotension in ewes. Anesthesiology 58: 245–249

Subject Index